F-Gen

foreign statistical documents

hoover institution bibliographical series: xxviii   edited by joyce ball

HOOVER INSTITUTION BIBLIOGRAPHICAL SERIES: XXVIII

foreign statistical documents

A Bibliography of General, International Trade, and Agricultural Statistics, Including Holdings of the Stanford University Libraries

Edited by Joyce Ball
and compiled by Roberta Gardella
of the Stanford University Library

The Hoover Institution on War, Revolution and Peace
Stanford University, Stanford, California 1967

The Hoover Institution on War, Revolution and Peace, founded at Stanford University in 1919 by the late President Herbert Hoover, is a center for advanced study and research on public and international affairs in the twentieth century. The views expressed in its publications are entirely those of the authors and do not necessarily reflect the views of the Hoover Institution.

© 1967 by the Board of Trustees of the
Leland Stanford Junior University
All rights reserved
Library of Congress Catalog Card Number: 67-14234
Printed in the United States of America

PREFACE

Stanford's libraries house exceptionally strong collections in official documents, particularly in the statistical publications represented in this bibliography. This strength is no accident: the Food Research Institute and the Hoover Institution on War, Revolution and Peace—both inspired and materially assisted by Herbert Hoover—have added an extraordinary dimension to the University's acquisitions interests and, consequently, to the collections of the libraries. Mrs. Joyce Ball and Mrs. Roberta Gardella, both members of the staff of the University Libraries throughout the compilation of the present bibliography, have by this work carried forward and added luster to the work of their predecessors, Mrs. Helen Farnsworth, Miss Minna Stillman and Miss Nina Almond.

The fundamental effort that has gone into this project has contributed greatly to the improvement of campus holdings through elimination of unnecessary duplication, consolidation of files, and closing of gaps. Beyond this, however, it is confidently expected that the bibliography presented herewith will prove useful within and without the University as a means of verifying citations, reflecting holdings, and guiding acquisition.

Rutherford D. Rogers
Director of University Libraries

INTRODUCTION

In 1946 it was necessary for all Stanford University libraries to reactivate arrangements for acquiring foreign statistical documents, many of which had not been published during World War II. At the suggestion of Mrs. Helen Farnsworth, then chairman of its Library Committee, the Food Research Institute, with the cooperation of Miss Minna Stillman, head of the Document Division of the Stanford Library, and Miss Nina Almond, librarian of the Hoover Library on War, Revolution and Peace, prepared a list of the annual publications of general statistics, trade statistics, and agricultural statistics available in the three libraries, including a record of issues held. The listing, originally compiled as an indicator of necessary acquisitions, proved to be a useful reference and research tool.

In 1958 a more comprehensive survey was made that included monthly or quarterly publications related to the above annuals, and added annual reports of agriculture departments. This survey was made by Miss Stillman, then technically retired, and was jointly financed by the Stanford Library and the Food Research Institute.

In March 1964, Bruce F. Johnston and Rosamond Peirce of the Food Research Institute presented to the Committee on International Studies a request for funds to bring the 1958 survey up to date and to acquire statistical documents to supplement holdings listed by the survey. Funds were granted to support this project from several sources—the International Studies Fund, the Library Development Fund, the Jordan Fund, and general University funds. As Foreign Document Librarian of the Stanford Library, I undertook supervision of the project.

The present project continued over a period of twenty months and differed from the previous two in several respects. For example, at the same time that holdings of the libraries were being surveyed, sets were consolidated, and duplicates not actively used in more than one location were eliminated. An effort was made to acquire missing publications and material never held in any of the three libraries. And a program of binding was initiated to preserve all material in the best possible manner.

The new bibliography includes call numbers, which the previous list omitted. It also identifies titles *not* held by any of the three libraries, and thus is designed to serve as a guide for future acquisitions as well as to provide additional source information for researchers.

It must be remembered that the bibliography has pragmatic limits; it is only as comprehensive as was deemed to be both practical for general use at Stanford and possible within the restrictions imposed by factors of funds and time. Efforts were concentrated mainly on publications using a Western European language either as the first or second language; and only very general material in other languages was included if such material was held at Stanford. As is typical of similar works, this bibliography was outdated before it reached the publisher. Because the document collection continues to grow, people searching for statistics not listed here should further check library catalogs and holdings records.

The editor is grateful to Roberta Gardella, who became much more than a compiler during the course of the project, she was dynamo and conscience both, and without her efforts much less would have been accomplished in the time allotted.

Mrs. Joan Lowcock, employed by the Government Document Division during most of the project, contributed a much-needed sense of calm to the project and its employees. However, this was not her only contribution, for she worked tirelessly with mountains of material, carried on extensive correspondence with persons in all countries of the world, and at the same time created records and procedures that will assure maintenance of the collection in the future.

This was indeed a cooperative project. Mr. Charles Milford of the Food Research Institute Library, Mr. Joseph Bingaman of the Hoover Institution, and Miss Gladys Doolittle of Stanford's Catalog Department gave generously of their time and patience. Much of the work was absorbed by regular staff members of the three institutions involved, all of whom accepted increased work loads graciously. Miss Susan Evans, Mr. Peter Cline, and Mr. Steve Lewis, Stanford students assigned to the project, contributed brainpower, initiative, and muscle power without which the less esoteric aspects of the work would have gone undone.

The impetus for the project came from the staff of the Food Research Institute, and their belief in the merit of such an undertaking sustained the project personnel through many seemingly impossible situations. The confidence expressed by the administration of the Stanford Library in their support of the project to its completion was a source of great personal gratification.

Finally, the editor and compiler share with scholars using the bibliography an eternal debt to Miss Lise Hofmann, who prepared the typescript.

Joyce Ball

Reno, Nevada
August 1, 1966

GUIDE TO THE USE OF THE BIBLIOGRAPHY

Material is listed alphabetically by country, then by type of publication as listed below, then in reverse chronological order with the latest material first:

SA   General statistics, annuals

SB   General statistics, bulletins (monthly or quarterly)

CA   Trade statistics, annuals

CB   Trade statistics, bulletins (monthly or quarterly)

AA   Agriculture statistics, annuals

AB   Agriculture statistics, bulletins (monthly or quarterly)

AR   Annual reports of departments of agriculture

Holdings are for the Stanford Library unless specified HL for Hoover Library or FRI for Food Research Institute Library. Call numbers were given for both of these collections; however, both have large uncataloged collections and the absence of a call number indicates the material is in one of these.

Symbols were used as follows:

-     After the initial date in the entry indicates the serial is still being published.

+   After holdings indicates the holding library has commitments to receive future publications as issued.

//   Ceased publication.

*   After a closing date indicates no further publications were found to be issued but there was no official statement located concerning discontinuation.

**   Entries from Gregory's List of the Serial Publications of Foreign Governments, 1815-1931 (New York, H. W. Wilson Company, 1932) were included for historical continuation but not held in one of the Libraries.

A

ADEN

see also FEDERATION OF SOUTH ARABIA

SA GREAT BRITAIN. Colonial Office.
  Annual report on Aden and Aden Protectorate. 1946-
  Previously issued in its numbered series, Colonial reports--annual (325.342 G787) no. 1883:1937-- no. 1936:1938. Suspended 1940-46.

  1946-58+   325.342
          G7871Ad

SA ADEN.
  The Aden blue book. 1937.
  Only one issue published; publication suspended for the duration of the war and not resumed.

SB ADEN.
  No general statistical bulletin available. The Aden Colony gazette has some data.

CA ADEN. Port Trust.
  Administration report.
  Each has a statistical appendix.

  1948/49-1964/65+ HE 560
          A2A3

CA ADEN.
  Trade and navigation report. 1867/68- 1953-54//
  Issued -1907/08 and 1914/15-1938 by the Trade Registration Office of Aden.
  Superseded by the Statement of external trade, issued by the Federation of South Arabia.

  1941-47, 1952-53, 1953-54 382.533
             A232

AFGHANISTAN

CA AFGHANISTAN. Ministry of Commerce. Statistical Department.
  A summary of Afghanistan's foreign trade.

  1962/63-1964/65 HF259
          A4A5

CB AFGHANISTAN. Ministry of Commerce.
  Exports of merchandise from Afghanistan. 1961- Monthly.

CB AFGHANISTAN. Ministry of Commerce.
  Imports of merchandise to Afghanistan. 1961- Monthly.

ALBANIA

SA ALBANIA. Drejtoria e Statistikës.
  Vjetari statistikor i R.P.SH. 1958-
  Translation of supplements only.

  1960, 1963-64+   314.974
  1963, English supplement A326
   Part I and II
  1964, English supplement

SA ALBANIA.
  Formularët e statistikës së përgjith.
  General statistics. 1924-26 (3 vols.)

SB ALBANIA.
  No general statistical bulletin available. Occasional statistical tables in Fletorja zyrtare, the official Albanian gazette.

CA ALBANIA.
  Statistika tregtare e vjetit.... Trade statistics for the year. 1921-27 (7 vols.)

CA **ALBANIA. Drejtoria e Kontabilitetit të Pergjitheshem.
  Statistika tregtare e importacjon-it e eksportacjon-it. Trade statistics. 1921-22(?)

ALGERIA

SA   ALGERIA. Sous-direction des Statistiques.
       Annuaire statistique de l'Algérie. 1926-
       1926-27 numbered Nouv. sér., v. 1-2; 1939/47-1962, Nouv. sér., v. 1-14; 1963-64-- Nouv. sér., v. 1-
       Supersedes Statistique générale de l'Algérie, issued by Service de la Statistique Générale.

       1933-34, 1936, 1939-61,   HA2071
       1963-64+   A32

SA   GREAT BRITAIN. Dept. of Overseas
CA   Trade.
       Report on economic conditions in Algeria, Tunisia and Tripolitania .... [1921-36.]
       Statistics for Algeria, 1924/25-1926/27, are found in the Survey of economic conditions in Morocco, 1924/25-1926/27. (Gt. Brit. Dept. of Overseas Trade)
       Title varies.

       1921-23, 1929-30, 1932,   380.005
       1935, 1936   G786

SA   ALGERIA. Service de la Statistique Générale.
       Statistique générale de l'Algérie. 1867/72-1925//

SA   **ALGERIA.
       Tableau de la situation des établissements français dans l'Algérie. 1837-1865/66//
       Continued as Statistique générale de l'Algérie.

SA   ALGERIA. Sous-direction des Statistiques.
       Bulletin mensuel de statistique générale. no. 1, Jan./Feb. 1952-
       Vols. for 1952-July 1962 issued by Service de Statistique Générale.

       no. 1-81; Jan./Feb. 1952-   316.5
           Jan. 1961   A395bm
       n. s. 1961-no. 7, Oct. 1965+
       wanting: no. 49-58; n. s.
           1962, no. 3

SB   ALGERIA. Service de Statistique Générale.
       Bulletin de statistique générale. v. 1, no. 1-v. 13, no. 3; Jan. 1949-Sept. 1961//
       Issued quarterly with monthly supplements Jan. 1949-Dec. 1959.

       v. 1, no. 1-v. 13, no. 3   316.5
                                                A395b

CA   ALGERIA. Service National des Douanes.
       ... Documents statistiques réunis par l'administration des douanes sur le commerce de l'Algérie.
       At head of title, 1901-1921, République Française. Gouvernement Général de l'Algérie. Direction des Douanes de l'Algérie.

       1909-14, 1919, 1921, 1955,   380.0965
       1960, 1961 (2 vols.)+   A395

CB   ALGERIA. Ministère de l'Economie Nationale.
       Statistique mensuelle des échanges commerciaux.

       Jan. 1964-July 1964+   HF268
                                                A4A35
                                                f

CB   ALGERIA. Ministère de l'Economie Nationale. Douanes Algériennes.
       Bulletin comparatif semestriel du mouvement commercial et maritime de l'Algérie.

       Jan./June 1962-Jan./June   HF268
       1963+   A4A3
                                                   f

AA   **ALGERIA. Service Central de Statistique Générale.
       Statistique agricole de l'Algérie. [1901/23(?)]

AB   ALGERIA. Ministère de l'Agriculture et de la Réforme Agraire.
       L'Algérie agricole. Algiers. 1, 1964-
       Supersedes Agriculture algérienne. Monthly(?)

AB **ALGERIA.
  Bulletin agricole de l'Algérie, Tunisie, Maroc. 1-35, 1895-1929//
  1-19, 1895-1913, as Bulletin agricole de l'Algérie et de la Tunisie.

ANGOLA

SA   ANGOLA. Repartição de Estatística
        Geral.
     Anuário estatístico. 1933-
        At head of title, 1933- : Colónia de Angola. Repartição Central de Estatística Geral; later, Colónia de Angola. Repartição Técnica de Estatística Geral.

     1949-63+   HA2211
                                 A3

SA   GREAT BRITAIN. Dept. of Overseas
CA   Trade.
     ... Economic conditions in Angola (Portuguese West Africa) .... [1923-29.]
        Notes on the financial situation in Angola for 1926, 1928, and 1930 are found with the Economic conditions in Portugal, Mar. 1926, Mar. 1928, and Mar. 1930, respectively. (Gt. Brit. Dept. of Overseas Trade)
        Title varies.

     1923, 1925, 1929   380.005
                                 G786

SA   **PORTUGAL. Ministério das Colónias.
     Relatórios e informações. 1, 1917//
        Contains material on Mozambique.

SB   ANGOLA. Repartição de Estatística
        Geral.
     Boletim mensal de estatística. 1945-

     v. 7, no. 1, Jan. 1951-   316.73
        v. 21, no. 8, Aug. 1965+ A592b

SB   ANGOLA. Repartição Técnica de Estatística Geral.
     Boletim estatístico. Trimestral.
     ano 1--1933, ano 2--1934, ano 3--1942, ano 4--1943.

CA   ANGOLA. Repartição de Estatística
        Geral.
     Comércio externo. Commerce extérieur.

     1945-63+   382.673
                                 A592

CA   ANGOLA. Repartição de Estatística
        Geral.
     Estatística das contribuições e impostos.

     1951-63+   336.206673
                                 A592

CB   **ANGOLA. Direcção dos Serviços Aduaneiros.
     Estatística comercial. Boletim trimestral. 1923-?

AB   **ANGOLA. Direcção dos Serviços da
        Agricultura.
     Boletim de agricultura.
        1908-09 as Boletim de agricultura, pecuaria e fomento.
        v. 1, no. 1-12: Nov. 1908-Oct. 1909; sér. 2, no. 1-6: 1912; sér. 3, no. 1: Apr. 1915; sér. 4, no. 1-8: Jan.-Aug. 1919; sér. 5, no. 1-2: 1920//

ANTIGUA

SA   GREAT BRITAIN. Colonial Office.
     Antigua; report. 1955/56-
        Continues in part its Report on the Leeward Islands, 1940-54 (325.342 G787ILe)

     1955/56-1961/62+   325.342
                                 G7871An

SA   **ANTIGUA.
     Blue book. 1821-87//
        1821-25 includes Montserrat.
        Continued in Leeward Islands. Blue book.

CA   ANTIGUA. Ministry of Trade and Production.
     Trade report. Annual.

AR   ANTIGUA. Dept. of Agriculture, Lands, Marketing and Credit.
    Report. 1915/16-
    Issued by the Dept. under earlier names.
    Supersedes Antigua. Botanic Station. Reports on the Botanic Station, economic experiments, and agricultural education, Antigua.

1932, 1934/38, 1951/52, 1954, 1958-59, 1960-62+	630.972972 A629

ARGENTINE REPUBLIC

SA   ARGENTINE REPUBLIC. Dirección Nacional de Estadística y Censos.
    Anuario estadístico de la República Argentina. 1948-57.
    1948 in 2 vols.; 1949/50 in 3 vols., I. Compendio. II. Comercio. III. Estadística industrial; 1957 with retrospective figures to 1937.
    No longer published.

1948, 1949/50	HA941 A32
FRI 1949/50, 1957	

SA / CA   GREAT BRITAIN. Dept. of Overseas Trade.
    ... Economic conditions in the Argentine Republic .... [1919-38.]
    1919 issued in the series of Parliamentary Papers as Papers by command. Cmd. 895.
    Title varies.

1919, 1921-25, 1927-31, 1933, 1935-38	380.005 G786

SB   ARGENTINE REPUBLIC. Dirección Nacional de Estadística y Censos.
    Boletín de estadística. Jan. 1956-
    Issued monthly 1956-62; quarterly 1963- .

Jan. 1956-July/Sept. 1965+ wanting: Nov., Dec. 1958	330.982 A693
FRI Jan. 1956-Oct./Dec. 1963	

SB   ARGENTINE REPUBLIC. Dirección Nacional de Estadística y Censos.
    Síntesis estadística mensual de la República Argentina. Jan. 1947-Dec. 1955//

v. 1-2, 4-9; 1947-48, 1950-55 wanting: v. 2: 5, 1948; v. 5: 12, 1951	318.2 A692s
FRI Jan. 1947-Dec. 1955	

CA   ARGENTINE REPUBLIC. Dirección Nacional de Estadística y Censos.
    Comercio exterior argentino. Anual. 1951-
    1951-54 in 1 vol. published 1959.
    1961- issued in 2 vols. each year.

1951-64+	382.82 A692ca

CA   ARGENTINE REPUBLIC. Dirección Nacional del Servicio Estadístico.
    Intercambio comercial argentino por paises, 1951-54.
    Supplement to Anuario estadístico. Issued as its Informe C 22-

1951-54	382.82 A692i

CA   ARGENTINE REPUBLIC. Dirección General de Estadística.
    Anuario estadístico de la República Argentina; comercio exterior. 1915-47//
    Superseded by Dirección Nacional de Estadística y Censos. Anuario estadístico ... and later its Comercio exterior. Informe C.

1917-26, 1930-35, 1938-49	382.82 A619a

CA **ARGENTINE REPUBLIC. Ministerio de Hacienda.
    Estadística del comercio y de la navegación. 1-13, 1880-92//
    1880-81 as Estadística del comercio exterior y de la navegación interior y exterior. Continues its Estadística general del comercio esterior. Continued as Dirección General de Estadística. Anuario.

CA **ARGENTINE REPUBLIC. Ministerio de
     Hacienda.
   Estadística general del comercio
 esterior. 1870-80//
   Continues its Estadística de la aduana
 ... 1870 as Estadística de las aduanas de
 la República Argentina; 1871-74, Estadística general del comercio esterior;
 1876-79, Cuadro general del comercio
 exterior. Continued by its Estadística
 del comercio exterior y de la navegación
 interior y exterior.

CA **ARGENTINE REPUBLIC. Ministerio de
     Hacienda.
   Estadística de la aduana de Buenos
 Aires. 1-9, 1861-69//
   Continued as its Estadística de las
 aduanas ..., later Estadística general
 del comercio esterior.

CB   ARGENTINE REPUBLIC. Dirección
     Nacional de Estadística y Censos.
   Comercio exterior. Informe C.
 1950/51-
   Issues cover irregular periods of time
 1950/51-1954.
   Sept. 1956-   Monthly.

   1950/51, 1952/53, 1954,   382.82
     Sept. 1956-Oct. 1965+   A692c

CB   ARGENTINE REPUBLIC. Dirección
     Nacional de Estadística y Censos.
   Intercambio comercial argentino con
 los paises de la A.L.A.L.C. 1, 1962-

   1-14, 1962-65+   382.82
                                  A692ic
   FRI 1-15, 1962-65+

CB   ARGENTINE REPUBLIC. Dirección
     General de Estadística.
   El comercio exterior argentino. 1,
 1882-1944/45//
   Published at various frequencies in
 earlier years, later published in two
 issues yearly: En los primeros semes-
 tres de ... y ... and Estadísticas eco-
 nómicas retrospectivas.

   no. 168, 176, 185, 189-   382.82
     200, 203, 206-209,   A691c
     211, 213-222, 224-
     228, 230-236;
     1916/17-1944/45

AA   ARGENTINE REPUBLIC. Dirección de
     Economía Rural y Estadística.
   Anuario agropecuario. 1906/07(?)-

   FRI 1917-18, 1925-26,
          1932-35

AA **ARGENTINE REPUBLIC. Dirección de
     Economía Rural y Estadística Agrí-
     cola.
   Estadística agrícola. Statistique
 agricole. 1902/04-1917/18.

AB   ARGENTINE REPUBLIC. Dirección de
     Economía Rural Estadística.
   Boletín de estadística y economía
 agropecuaria. 1959-
   Ceased temporarily Apr. 1961.

   FRI 1959: 1-4, 7-12;
          1960: 1-4

AB   ARGENTINE REPUBLIC. Dirección de
     Economía Rural y Estadística.
   Boletín estadístico.
   Monthly, 1928, 1931-   ; quarterly,
 1929-June 1930; semiannual, July-Dec.
 1930.
   Title varies: Boletín mensual de
 estadística agrícola; 1927-36, 1943,
 Boletín mensual de estadística agropecuaria; 1937-42, Boletín estadística
 agropecuaria; 1944, Estadística; boletín
 mensual; 1945-   Boletín estadístico.

   no. 29-33, 35-47; 1927-32,   630.682
     1934-46   A690b
   wanting: Jan.-May, July-
     Aug. 1931; Nov.-Dec.
     1936

   FRI [1922-46]

AR   ARGENTINE REPUBLIC. Ministerio de
     Agricultura.
   Memoria.

   1916, 1923-24, 1928,   630.682
     1934, 1937-44,   A691m
     1946-48

ARUBA

see NETHERLANDS ANTILLES

AUSTRALIA

SA AUSTRALIA. Bureau of Census and
    Statistics.
    Official year book of the Commonwealth of Australia. 1, 1908-

    1908-65+   HA3001
                                B5
    FRI   Current two years
          only.

SB AUSTRALIA. Bureau of Census and
    Statistics.
    Quarterly summary of Australian statistics. 1, Jan. 1912-
    Issued monthly Jan. 1912-Dec. 1917.

    no. 1-258; 1912-65+   319.4
                                 A93q
    FRI   Current two years
          only.

CA AUSTRALIA. Bureau of Census and
    Statistics.
    Oversea trade. Bulletin. no. [1], 1906-
    Formerly published by the Dept. of Trade and Customs with title: Annual statement of the trade of the Commonwealth of Australia with the United Kingdom, British possessions, and foreign countries.
    Title varies.

    1906-1938/39, 1944/45-   382.94
    1964/65+   A938

CB AUSTRALIA. Bureau of Census and
    Statistics.
    Bulletin of oversea trade statistics. Nov. 1954-

    Nov. 1954-Dec. 1963+   382.94
                                 A938t
    FRI   Current two years
          only.

CB AUSTRALIA. Bureau of Census and
    Statistics.
    ... Trade, shipping, oversea migration, and finance of the Commonwealth of Australia.... Bulletin. no. [1]-60, 1907-11.
    Superseded by its Monthly [now Quarterly] summary of Australian statistics.

    1907-11   382.94
                                 A938b

AA AUSTRALIA. Bureau of Census and
    Statistics.
    Statistical bulletin: rural land use and crop production.

    FRI   1960/61-1963/64

AA AUSTRALIA. Bureau of Census and
    Statistics.
    Rural industries. Bulletin. no. 1, 1962/63-
    Supersedes its Primary industries. Bulletin, pt. 1.

    1, 1962/63+   HD9018
                                 A7A3

AA AUSTRALIA. Bureau of Census and
    Statistics.
    Primary industries. Bulletin. no. 44-56; 1949/50-1961/62//
    Issued in two parts, 1950/51-1961/62: pt. 1, Rural industries; pt. 2, Non-rural industries and value of production.
    Superseded by its Rural industries. Bulletin and Non-rural primary industries ....

    1949/50-1961/62   330.994
                                 A931
    FRI   1949/50-1958/59

AA AUSTRALIA. Bureau of Census and
    Statistics.
    Production. Bulletin. 1901-06--1948/49//
    Issued in parts 1936/37-1948/49.
    Superseded by the bureau's Primary industries and Secondary industries.

    1901-1948/49   330.994
                                 A93
    FRI   1911/12-1948/49

AA AUSTRALIA. Bureau of Census and
    Statistics.
    Secondary industries. Bulletin.
    no. 44, 1949/50-
        Continues volume numbering of its
    Production, which it supersedes in part.

    1949/50-1962/63+   330.994
                                    A932
    FRI 1949/50-1960/61

AB AUSTRALIA. Bureau of Agricultural
        Economics.
    Quarterly review of agricultural eco-
    nomics. 1, Jan. 1948-

    Jan. 1948-Jan. 1966+   330.9106
    wanting: v. 11, no. 1   A938
        (1959)

    FRI 1948-Jan. 1966+

AUSTRIA

SA AUSTRIA. Bundespressedienst.
    Oesterreichisches Jahrbuch ....
    1918/19-
        None published 1937-44.
        Published also in English.

    1949-64+   HC261
    wanting: 1950   O8

    English ed., 1929-31   314.36
                                    A93

    HL 1918/19-1920, 1926,   DB1
        1930, 1932-36,   A2
        1945/46, 1948-49,
        1961

SA AUSTRIA. Statistisches Zentralamt.
    Statistisches Handbuch für die
    Republik Oesterreich. 1920-
        Issued 1918-37 by Bundesamt für
    Statistik.
        Suspended 1938-49.

    1920-24, v. 1-5; 1927-37,   HA1171
        v. 8-17; 1950-65, neue   A3
        Folge v. 1-16

    HL 1923-25, 1927-32,   HA1171
        1935-37, 1950-57,   A938
        1960

SA AUSTRIA. Statistisches Landesamt.
    Statistisches Jahrbuch für Oester-
    reich. 1938-

    1938   314.36
                                    A939j

    HL 1938   HA1171
                                    A9381

SA AUSTRIA. Statistische Zentralkommis-
        sion.
    Oesterreichisches statistisches Hand-
    buch für die im Reichsrathe vertretenen
    Königreiche und Länder 1-35,
    1882-1916/17//

    Jahrg. 5, 11, 13-14,   314.36
        29-31, 34, 35;   A938
        1886, 1892, 1894-95,
        1910-12, 1915,
        1916/17

    HL 1911, 1916/17   HA1171
                                    A939

SA AUSTRIA. Statistische Zentralkommis-
        sion.
    Oesterreichische Statistik. v. 1-93,
    1880-1910. Neue Folge v. 1-18, 1910-18.

    v. 1-93, 1882-1910   314.36
    wanting: 3, 5, 9, 31-35,   A938o
        48-49, 51-54, 57-58,
        60, 62-64
    n. F. v. 1-18
    wanting: several parts of
        vols.

SA AUSTRIA. Statistische Zentralkommis-
        sion.
    Statistisches Jahrbuch der österreichi-
    sches Monarchie. 1863-81//

    1863, 1866, 1867   314.36
                                    A938sj

SA GREAT BRITAIN. Dept. of Overseas
CA   Trade.
    ... Report on the financial, commer-
    cial and industrial situation of Austria
    .... [1921-37.]
        Title varies.

    1921-27, 1929, 1933,   380.005
        1935, 1937   G786

SB AUSTRIA. Statistisches Zentralamt.
    Statistische Nachrichten. Apr. 1923-
    Suspended 1939-Sept. 1946; resumed
    publication with Neue Folge, Bd. 1,
    Sept. 25, 1946.
    Issued Apr. 1923-Feb. 1938 by
    Bundesamt für Statistik; Apr.-Dec. 1938,
    by Statistisches Landesamt.
    Supersedes Mitteilungen of the Bundes-
    amt für Statistik.

    1923-38, Sept. 1946-63+   314.36
                                   A937n
                                      f
    FRI   Current two years
          only.

SB AUSTRIA. Statistisches Zentralamt.
    ... Beiträge zur österreichischen
    Statistik. 1, 1946-
    See also AA for FRI holdings on
    Ergebnisse der Landwirtschaftlichen
    Statistik, which is part of the Beiträge
    ... series.

    no. 1-116, 1946-65+   314.36
                                   A939

SB AUSTRIA. Statistische Zentralkommis-
    sion.
    Beiträge zur Statistik.
    Merged with Statistische Nachrichten,
    Apr. 1923.

    1919: Heft 1; 1920: Heft   314.36
    3-4; 1921: Heft 6, 7,   A938b
    9-11

SB AUSTRIA. Statistische Zentralkommis-
    sion.
    Statistische Monatsschrift. 1875-
    1917; 1919-21//
    Vol. 22, Jahrg. 1896- numbered
    also Neue Folge I-
    Supplements issued with earlier vols.

    3. Folge, v. 1-3, Heft 6;   314.36
    1919-21   A938sm

CA AUSTRIA. Statistisches Zentralamt.
CB Der Aussenhandel Oesterreichs.
    1919-
    Issued 1919-46 by Bundesministerium
    für Handel und Verkehr.
    Continues Austria. Handelsstatisti-
    scher Dienst. Statistik des auswärtigen
    Handels des Vertragszollgebietes der

beiden Staaten der Oesterr.-Ungar.
Monarchie, 1891-1917 (382.436 A938st).
    Title varies: 1919-29, Statistik des
auswärtigen Handels Oesterreichs;
1930-63, Statistik des Aussenhandels
Oesterreichs. Vols. for 1964- issued
in 3 series.

    1919, 1923-28, 1948-65+   382.436
    (1919 Halbjahr only)   A938sta

CA AUSTRIA. Handelsstatistischer Dienst.
    Statistik des auswärtigen Handels des
    Vertragszollgebietes der beiden Staaten
    der Oesterr.-Ungar. Monarchie. 1891-
    1917//
    Several vols. each year.

    1891: 1-3; 1892: 2; 1893:   382.436
    2-3; 1894: 1; 1895: 1,   A938st
    3; 1898: 1, pt. 2; 1908:
    4; 1910-13; 1914: 1-3;
    1915: 1-4; 1916: 1-4;
    1917: 1-2

CB **AUSTRIA. Bundesministerium für Handel
    und Verkehr.
    Monatshefte der Statistik des Aussen-
    handels Oesterreichs. 1930-(?)
    Continues the quarterly ed. of Statistik
    des auswärtigen Handels.

CB **AUSTRIA. Bundesministerium für Handel
    und Verkehr.
    Statistik des auswärtigen Handels ....
    Quarterly. 1926-29//
    Continued as Monatshefte der Statistik
    des Aussenhandels Oesterreichs.

    1926: 1; 1927: 1-4; 1928:   382.436
    1-3. Bound with   A938sta

AA AUSTRIA. Statistisches Zentralamt.
AB Ergebnisse der landwirtschaftlichen
    Statistik.
    These appear in the numbered series
    Beiträge zur österreichischen Statistik,
    which also includes other reports.

    FRI 1946, 1953-56,
         1960-63, 1965+

AA   AUSTRIA. Bundesministerium für Land- und Forstwirtschaft.
     Statistik der Ernte.
     Vols. for 1910-17 issued by Ackerbauministerium; 1918-19, by Staatsamt für Land- und Forstwirtschaft.
     Title varies: 1910-13, Anbau und Ernte der wichtigsten Körnerfrüchte; 1914-(?), Anbauflächen und Ernteergebnisse.

     1910-21, 1925-34, 1936   630.6436
                                  A938a

AA   AUSTRIA. Ackerbauministerium.
     Statistisches Jahrbuch. 1874-1913//
     Issued in parts: 1. Statistik der Ernte; 2. Bergwerksbetrieb Oesterreichs; 3. Forst- und Jagdstatistik (quinquennial).

     1895-1913 (pt. 1 only,   630.6436
       except for 1910,   A938
       pt. 1, 3)

B

BAHAMAS

SA   GREAT BRITAIN. Colonial Office.
     Annual report on Bahama Islands.
     Previously issued in its numbered series, Colonial reports--annual (325.342 G787) no. 1095: 1920-21-- no. 1901: 1938. Suspended 1940-46.

     1946--1962-63+   325.342
                                  G7871Bh

SA   BAHAMAS.
     Blue book. 1821-1939.

     1914-16, 1921-29   354.7296
                                  B151b

CA   BAHAMAS. Customs Dept.
     Statistical return. 1958-
     1964 has title Administrative report and trade return.

     1958, 1964   HF147
       (Each has figures for   A3
       four preceding years)   f

CA   BAHAMAS.
     [Final returns, imports-exports]
     Nassau. 1957-

CA **BAHAMAS. Customs Dept.
     Report [on the customs, revenue, trade and shipping]. 1918-

AB **BAHAMAS. Dept. of Agriculture.
     Bulletin. v. 1-6, no. 1; 1906-Mar. 1911//

AR   BAHAMAS. Agricultural and Marine Products Board.
     Annual report. 1904/06-

BAHREIN

SA   BAHREIN.
     Annual report. 1949/50-

     1949/50-1963+   354.539
     wanting: 1962   B151

SA   GREAT BRITAIN. Dept. of Overseas
CA   Trade.
     ... Economic conditions in the Persian Gulf ... Report .... [1929-36.]

     1929-34, 1936   380.005
                                  G786

BARBADOS

SA   BARBADOS. Statistical Service.
     Abstract of statistics. 1, 1956-

     1, 1956; 2, 1957; 3, 1960;   317.298
     4, 1963 (Suppl. 1964)   B228

SA   GREAT BRITAIN. Colonial Office.
     Annual report on Barbados. 1946-
     Previously issued in its numbered series, Colonial reports--annual (325.342 G787) no. 1072: 1919-20-- no. 1913: 1938-39. Suspended 1940-46.

     1947-61+   325.342
                                  G7871Ba

SA   BARBADOS.
       Blue book. 1821-

     1914/15-1918/19,   354.7298
     1945-46   B228

SB   BARBADOS. Statistical Service.
       Quarterly digest of statistics. 1, Dec. 1956-

     no. 29, Mar. 1964-no. 31,   HA865
     Sept. 1964+   A312

CA   BARBADOS. Statistical Service.
       Overseas trade. 1957-
       Includes information formerly issued in Customs Dept. Report of the comptroller of customs.

     1957-63+   382.7298
                               B229
                               f

CA   BARBADOS. Customs Dept.
       Administration report of the comptroller of customs. 1956-
       Continues in part its Report of the comptroller of customs.

     1956-63+   382.7298
                               B228ar

CA   BARBADOS. Customs Dept.
       Report of the comptroller of customs on the customs revenue, trade, shipping and excise of the island.
       Continues Report on the trade statistics, 1896/1910.
       Continued by its Administration report of the ... and by Barbados. Statistical Service. Overseas trade.
       Title varies: 1914- Report on the Customs Dept.

     1914-18, 1949-55   382.7298
                               B228
                               f

CB   BARBADOS. Statistical Service.
       Quarterly overseas trade report.
       "All figures in this report are preliminary and are subject to revision and adjustment at the end of the year."

     Current year only.   Gov't. Doc. Div.

AR   BARBADOS. Dept. of Science and Agriculture.
       Report. 1898/1907-(?)

BASUTOLAND

SA   GREAT BRITAIN. Office of Commonwealth Relations.
       Annual report on Basutoland. 1946-
       Previously issued in its numbered series, Colonial reports--annual (325.342   G787) no. 1085: 1920-21-- no. 1898: 1938. Suspended 1940-46.

     1946-63+   325.342
                               G7871Bs

SA   **BASUTOLAND.
       Blue book. 1913/14-(?)

AR   BASUTOLAND. Dept. of Agriculture.
       Report. Annual.

     1952-62+   630.9684
                               B327
     FRI   Current two years only.

BECHUANALAND

see also SOUTH AFRICA, Official year book of the Union and of Basutoland, Bechuanaland Protectorate and Swaziland

SA   GREAT BRITAIN. Colonial Office.
       Annual report on the Bechuanaland Protectorate. 1946-
       Previously issued in its numbered series, Colonial reports--annual (325.342   G787) no. 1083: 1920-21-- no. 1911: 1938. Suspended 1940-46.

     1946-64+   325.342
                               G7871Bch

SA   **BECHUANALAND.
       Blue book. 1913/14-(?)

AR BECHUANALAND. Dept. of Agriculture.
　　Annual report. 1951-
　　　No reports issued for 1959-61. Figures for these years to be found in Gt. Brit. Colonial Office. Annual report on the Bechuanaland Protectorate.

　　FRI　1962/63+

BELGIUM

SA　BELGIUM. Institut National de Statistique.
　　Annuaire statistique de la Belgique. 1870-
　　　1870-1955, Annuaire statistique de la Belgique et du Congo belge.

　　1878-1964+　　　　　　　314.93
　　wanting: 1879-81, 1888,　B428
　　　1900, 1903-05, 1912-13,
　　　1920-21, 1929-32,
　　　1937-39, 1947-48

SA　BELGIUM. Institut National de Statistique.
　　Annuaire statistique de poche.

　　1965+　　　　　　　　　HA1393
　　　　　　　　　　　　　　A32

SB　BELGIUM. Institut National de Statistique.
　　Bulletin de statistique. Sept. 1909-
　　Nouv. sér., 22e année, Jan. 1936-
　　Suspended 1914-19.
　　1910-35: title, Bulletin trimestriel.

　　Mar. 1910-Dec. 1943　　314.93
　　　(incomplete), 1944-65+　B428b

CA　BELGIUM. Institut National de Statistique.
　　Statistique annuelle du commerce extérieur de l'Union économique belgo-luxembourgeoise. Jaarstatistiek over de buitenlandse handel van de Belgisch-Luxemburgse Economische Unie. 1949/50-
　　　No longer published; see Benelux for later figures.

　　1949/50　　　　　　　　382.493
　　　　　　　　　　　　　　B428s
　　　　　　　　　　　　　　f

CA　BELGIUM. Ministère des Finances.
　　Tableau annuel du commerce avec les pays étrangers. 1831-1926//

　　1910-13, 1920-26　　　　382.493
　　　　　　　　　　　　　　B429t
　　　　　　　　　　　　　　f

CB　BELGIUM. Institut National de Statistique.
　　Bulletin mensuel du commerce extérieur de l'Union économique belgo-luxembourgeoise.
　　　Dec. issues give cumulated statistics for year.
　　　Issued by various agencies, under various titles.

　　1932, 1938, 1939, Sept.　382.493
　　　1948-Oct. 1965+　　　　B428

AA　BELGIUM. Institut National de Statistique.
　　Annuaire statistique agricole de la Belgique.

　　1946-62+　　　　　　　630.6493
　　wanting: 1955　　　　　B429a

　　FRI　1946-60+

AA　BELGIUM. Institut National de Statistique.
　　Estimation de la production agricole; rendement des récoltes.

　　FRI　1928/29, 1941/42-
　　　　1946/47, 1951/52-
　　　　1962/63+

AA　BELGIUM. Institut National de Statistique.
　　Recensement agricole et horticole; résultats définitifs.

　　FRI　1942-66+

AA **BELGIUM. Ministère de l'Agriculture.
　　Statistique agricole. 1911-13//

AB   BELGIUM. Ministère de l'Agriculture.
        Revue de l'agriculture. Monthly.
      Contains statistical section.

      v. 2, 1949-65+   630.5
      wanting: Feb. 1957   R456

AB   BELGIUM. Ministère de l'Agriculture et
        des Travaux Publics.
        Bulletin de l'agriculture et de l'horti-
      culture. t. 1-3, no. 12: 1911-14.

      t. 1: no. 1, 5-12; t. 2:   630.6493
        no. 1-11; t. 3: no. 1-12   B428

AB **BELGIUM. Ministère de l'Agriculture.
        Bulletin de l'administration de l'agri-
      culture. v. 1-4, no. 6, 1908-June 1911//
      Continues Bulletin de l'agriculture.
      Continued as Bulletin de l'agriculture et
      de l'horticulture.

AB **BELGIUM. Ministère de l'Agriculture.
        Bulletin de l'agriculture 1-23,
      1885-1907//
         Continued as Bulletin de l'administra-
      tion de l'agriculture.

BERMUDA ISLANDS

SA   GREAT BRITAIN. Colonial Office.
        Annual report on Bermuda. 1946-
      Previously issued in its numbered
      series, Colonial reports--annual
      (325.342 G787) no. 1077: 1920-
      no. 1899: 1938. Suspended 1940-46.

      1946-1961/62+   325.342
                                   G7871Be

SA   BERMUDA ISLANDS.
        Blue book. 1821-

      1944, 1946-48   354.7299
                                   B517
                                   f

SA   GREAT BRITAIN. Dept. of Overseas
CA   Trade.
        ... Report on economic and commer-
      cial conditions in the British West Indies,
      British Guiana, British Honduras, and
      Bermuda .... 1921-37.

      1921-22, 1924-25, 1928,   380.005
        1930, 1932-34, 1937   G786

CA   BERMUDA ISLANDS.
        Imports and exports.

      1952-64+   382.7299
                                   B516

AR   BERMUDA ISLANDS. Dept. of Agricul-
        ture.
        Report. 1913-

      1947-64+   630.67299
      wanting: 1963   B516

BOHEMIA

SA   BOHEMIA. Office de Statistique.
        Zpravy zemského statistického uradu
      kralovstvi ceskeho. 1899-1918.

      1: 2; 2: 2; 7: 1; 10: 1-2;   314.371
        11: 1-2; 12: 2 and apx.;   B676
        13; 14: 1-2 and apx.;
        15: 1-2 and apx.; 16:
        1-2; 17: 1-2; 19: 1; 20:
        1-2; 21: 1-2; 22: 1-2;
        23: 2; 24: 1-2; 25: 2-3;
        26: 1; 1899-1918

SA   BOHEMIA. Zemský Statistický Urad.
        Statistisches Handbuch des König-
      reiches Böhmen. 1909-13.
         German ed. of its Statistická prirucka.
      Two vols.; v. 2, "II Ausgabe."

      1909-13 (v. 2 only)   314.371
                                     B676s

BOHEMIA AND MORAVIA

see also CZECHOSLOVAK REPUBLIC

SA   BOHEMIA AND MORAVIA (Protectorate, 1939-45). Statistisches Zentralamt.
Statistisches Jahrbuch für das Protektorat Böhmen und Mähren. v. 1-4, 1941-44//
    Continuation of Czechoslovak Republic. Státní Urad Statistický. Statistisches Jahrbuch der Cechoslovakischen Republik.

HL 1941, v. 1; 1942, v. 2	HA1191 A41

BOLIVIA

SA   BOLIVIA. Dirección de Estadística y Censos.
Extracto estadístico de Bolivia.
1936 in 3 sections.

SA   GREAT BRITAIN. Dept. of Overseas
CA   Trade.
... Economic conditions in Bolivia ... Report .... [1925-1931.]
    1925 has title: Report on the finance, trade and production of Bolivia ....
    Title varies.

1925, 1928, 1931	380.005 G786

SA   BOLIVIA. Dirección General de Estadística y Estudios Geográficos.
Anuario nacional estadístico y geográfico de Bolivia. 1-2, 1917-1919//
    None issued for 1918.

1917	318.22 B689a

SB   BOLIVIA. Dirección General de Estadística y Censos.
Boletín estadístico. 1901-
    Irregularly issued by various agencies.

no. 84, 1960; no. 87, 1963; no. 88, 1963	HA961 A35 f

SB   BOLIVIA. Dirección General de Estadística y Censos.
Suplemento estadístico. Bi-monthly. 1963-

FRI   Sept. 1963, Dec. 1963, Mar. 1964

SB   BOLIVIA. Dirección General de Estadística y Censos.
Revista mensual. July 1945-
Monthly: irregular.

CA   BOLIVIA. Dirección General de Estadística y Censos.
Importaciones.

1961	HF161 A4 f

CA   BOLIVIA. Dirección General de Estadística y Censos.
Balanza comercial de Bolivia, comercio exterior.

1950-60	382.822 B684f

CA   BOLIVIA. Dirección General de Estadística y Censos.
Comercio exterior. 1950/59-
    Supersedes Dirección General de Estadística. Comercio exterior: anuario.

CA   BOLIVIA. Dirección General de Estadística y Censos.
Comercio exterior: anuario. 1928-29--(?)
    Title varies: 1928-29, Comercio especial de Bolivia. 1932- Comercio exterior: anuario.
    Superseded by Dirección General de Estadística y Censos. Comercio exterior. 1950/59.

1928-29, 1932-37, 1942, 1948, 1949	382.822 B683

CA   BOLIVIA. Dirección General de
       Aduanas.
       Comercio especial de Bolivia. 1912-
1926/27(?)//

     1912   382.822
                                    B68

CA **BOLIVIA. Dirección General de
       Aduanas.
       Memoria. 1912/13-1916/17//
       Continued in its Comercio especial.

AA   BOLIVIA. Dirección General de Estadís-
       tica. Departamento de Estadísticas
       Agropecuarias.
       Resumen general de la República.
Estimaciónes agropecuarias. 1957/58-

AA   BOLIVIA. Dirección General de Estadís-
       tica.
       Estadística agropecuaria.

    FRI   1939-41

BRAZIL

SA   BRAZIL. Conselho Nacional de Estatís-
       tica.
       Anuário estatístico do Brasil.
1908/12-
      Quinquennial, 1908-12; annual, 1936-
1913-35 not published.

     1908/12, 1956-64+   HA971
                                       A32
    FRI   1939/40-1962,
          1964+

SA   BRAZIL. Instituto Brasileiro de Geogra-
       fia e Estatística.
       Sinopse estatístico do Brasil. 1934-
None published for 1941-45.
      Separata do Anuário estatístico do
Brasil.

     1938, 1946   318.1
                                   B821s

SA   GREAT BRITAIN. Dept. of Overseas
CA   Trade.
       ... Report on economic and commer-
cial conditions in Brazil.... [1921-38.]
      1919 issued in the series of Parlia-
mentary Papers as Parliament. Papers
by command. Cmd. 840.
      Title varies.

     1921-26, 1928-32, 1935,   380.005
       1937, 1938   G786

SB   BRAZIL. Conselho Nacional de Estatís-
       tica.
       Boletim estatístico. Jan.-June 1943-
1943-60 under agency: Instituto
Brasileiro de Geografia e Estatística.
      Issued monthly, later quarterly.

     July 1944-Jan./Mar. 1965+   318.1
                                          B821b
    FRI   1947-64+

SB   BRAZIL. Conselho Nacional de Estatís-
       tica.
       Revista brasileira de estatística.
Jan./Mar. 1940-
      Supersedes the Revista de economia e
estatística.

     1940-no. 100, 1964+   318.1
                                        R454
    FRI   [1940-64]

SB   BRAZIL. Serviço de Estatística Econô-
CB   mica e Financeira.
       Mensário estatístico. 1, July 1951-

     1, July 1951-168, June   318.1
       1965+   B8275m
    FRI   1951-June 1965+

CA   BRAZIL. Serviço de Estatística Econô-
       mica e Financeira.
       Comércio exterior do Brasil. 1901-
      Some vols. issued in 4 parts: pt. 1,
por paises, segundo os mercadorias;
pt. 2, por mercadorias, segundo os
paises; pt. 3, por portos, segundo os
mercadorias; pt. 4, valores em dolares.

     1913, 1915-18, 1919-23,   382.81
       1933-37, 1944-50,   B827
       1951/52-1964+

CA BRAZIL. Serviço de Estatística Econô-
      mica e Financeira.
    Foreign trade of Brazil according to
    the Standard International Trade Classifi-
    cation [SITC].

    1955/56-1959/60+   382.81
                               B827f
    FRI Current two years
         only.

CA BRAZIL. Serviço de Estatística Econô-
      mica e Financeira.
    Comércio exterior do Brasil. Resumo
    por mercadorias. 1900-
      Title varies: -1908, Boletim;
    --1908/09- Boletim da estatística
    comercial; 1912/13- Comércio exterior
    do Brasil.

    1920-23, 1928, 1932,   382.81
      1939, 1947-48, 1952   B827a

CB BRAZIL. Serviço de Estatística Econô-
      mica e Financeira.
    Estatística do comércio exterior.
      Issued quarterly with annual cumula-
    tions.
      Publication suspended 1962.

    1951-Sept. 1954, 1959   382.81
                               B827e

AA BRAZIL. Ministério da Agricultura.
    Anuário. 1928+

    1928-29   354.81
                               B84a

AA **BRAZIL. Serviço de Inspecção e
      Fomento Agricolas.
    Estatística agricola. 1922/23-(?)
    Name of dept. varies.

AB BRAZIL. Ministério da Agricultura.
    ... Boletim do Ministério da Agricul-
    tura, Indústria e Comércio. ano 1, Mar.
    1912-

    1912-1944/45   354.81
    wanting: many issues   B84b

AR BRAZIL. Ministério da Agricultura.
    Relatório [e anexos]. 1860-
      Name of ministry varies: 1860-92,
    Ministério da Agricultura, Comércio e
    Obras Públicas; 1912-26, Ministério da
    Agricultura, Indústria e Comércio.
      Reports for 1893-1909 were included
    in Relatório, Ministério da Viação e
    Obras Públicas, under a variant name
    (Ministério da Indústria, Viação e Obras
    Públicas).

    Relatório: 1884, 1886-88, 354.81
      1890-92, 1912-13,   B82
      1916-18, 1921-26,
      1938-39
    Anexos: 1885, 1887,
      1888, 1892

AR BRAZIL. Ministério da Viação e Obras
      Públicas.
    Relatório [e anexos]. 1893-
      Name of ministry varies: 1893-1909,
    Ministério da Indústria, Viação e Obras
    Públicas.
      Reports for 1893-1909 continue Rela-
    tório, Ministério da Agricultura, issued
    under a variant name (Ministério da
    Agricultura, Comércio e Obras Públicas).
    Beginning in 1910, Relatório, Ministério
    da Agricultura, issued separately.

    Relatório: 1893-99,   354.81
      1901-04, 1906-09   B83
    Anexos: 1893

BRITISH EAST AFRICA

see EAST AFRICA, KENYA, TANGANYIKA,
UGANDA, ZANZIBAR

BRITISH GUIANA

SA GREAT BRITAIN. Colonial Office.
    Annual report on British Guiana.
    1946-
      Previously issued in its numbered
    series, Colonial reports--annual
    (325.342 G787) no. 1109:1920-
    no. 1926:1938. Suspended 1940-46.

    1946-61+   325.342
                               G7871Br

SA   GREAT BRITAIN. Dept. of Overseas
CA   Trade.
        ... Report on economic and commer-
cial conditions in the British West Indies,
British Guiana, British Honduras, and
Bermuda.... [1921-37.]
      Title varies.

    1921, 1922, 1924-25,   380.005
      1928, 1930, 1932-34,   G786
      1937

SA   BRITISH GUIANA.
      Blue book. 1843-

    1910/11, 1917-24, 1926   318.421
                                       B358b

SB   BRITISH GUIANA. Statistical Bureau.
      Quarterly statistical digest.

CA   BRITISH GUIANA. Dept. of Customs and
      Excise.
      Account relating to external trade.
      Account for 1953 included in the
dept.'s Report of the comptroller of customs and excise.

    1955-61+   382.8421
    wanting: 1956, 1958   B862

CA   BRITISH GUIANA. Dept. of Customs and
      Excise.
      Report of the comptroller of customs
and excise. 1953-
      Supersedes British Guiana. Customs
Dept. Report of the comptroller of customs.

    1953-61+   336.26
                                         B863

CA   BRITISH GUIANA. Customs Dept.
      Report of the comptroller of customs.
1879-1952(?)//
      Superseded by British Guiana. Dept.
of Customs and Excise. Report of the
comptroller of customs and excise.

    1924-26, 1949-52   336.26
                                         B862

CA   BRITISH GUIANA.
AR   Administration reports.
      These reports contain some figures on
agriculture and commerce.

    1917-39   354.8421
    wanting: 1928, 1931   B861a

CB   BRITISH GUIANA. Statistical Bureau.
      Monthly account relating to external
trade.

    June 1964-Oct. 1964+   HF172
                                         B7A3

AR   BRITISH GUIANA. Dept. of Agriculture.
      Report of the director of agriculture.
1928-

    1951-62+   630.98421
                                       B862
                                       f

BRITISH HONDURAS

SA   BRITISH HONDURAS. Ministry of
      Finance.
      Annual abstract of statistics. 1961-

SA   GREAT BRITAIN. Colonial Office.
      Annual report on British Honduras.
1946-
      Previously issued in its numbered
series, Colonial reports--annual
(325.342 G787) no. 1110: 1920-
no. 1834: 1938. Suspended 1940-46.

    1946-1962/63+   325.342
                                     G7871Bd

HL   1959-61   F1443
                                       G787

SA   GREAT BRITAIN. Dept. of Overseas
CA   Trade.
      ... Report on economic and commer-
cial conditions in the British West Indies,
British Guiana, British Honduras, and
Bermuda.... [1921-37.]
      Title varies.

    1921, 1922, 1924-25, 1928,   380.005
      1930, 1932-34, 1937   G786

SA   BRITISH HONDURAS.
        Blue book. 1822-

SB   BRITISH HONDURAS. Public Relations
        Office.
        British Honduras monthly bulletin.
    Issued also in a Spanish ed.

CA   BRITISH HONDURAS. Customs Dept.
        Trade report. 1920-
        Title varies: 1920- Report on
imports, exports, excise and shipping.

    1954, 1955   382.7282
                                  B862
                                  f

AR   BRITISH HONDURAS. Agriculture Dept.
        Report.

    1953-64+   630.67282
    wanting: 1958, 1959   B862
                                  f

BRITISH NORTH BORNEO

AR **BRITISH NORTH BORNEO. Dept. of
        Agriculture.
        Annual report on agriculture.
    1925-(?)

BRITISH
SOLOMON ISLANDS

SA   GREAT BRITAIN. Colonial Office.
        Annual report on the British Solomon
Islands. 1946-
        Previously issued in its numbered
series, Colonial reports--annual
(325.342 G787) no. 1148: 1921-22--
no. 1908: 1938. Suspended 1940-46.

    1948-62+   325.342
                                  G7871So

SA   BRITISH SOLOMON ISLANDS.
        Blue book. 1920-

CA   BRITISH SOLOMON ISLANDS. Collector
        of Customs.
        Trade report, and annual report by the
Collector of Customs.
        Title varies: 1960-61, Trade statis-
tics.

    1960-64+   HF293
    wanting: 1962   S6A4

AR   BRITISH SOLMON ISLANDS. Dept. of
        Agriculture.
        Report.

    1960-64+   S400
    wanting: 1961   B7A3

BRITISH VIRGIN ISLANDS

see VIRGIN ISLANDS

BRUNEI

SA   GREAT BRITAIN. Colonial Office.
        Annual report on Brunei. 1946-
        Previously issued in its numbered
series, Colonial reports--annual
(325.342 G787) 1920-no. 1930: 1938.
Suspended 1940-46.

    1946--1961-62+   325.342
                                  G7871Bu

SA   BRUNEI.
        State of Brunei: annual report for ....
        Same as report listed here under
Gt. Brit. Colonial Office. Annual
report on Brunei.

CA   BRUNEI. Borneo (State) Dept. of Cus-
CB   toms and Excise.
        Statistics [of external trade].

    Jan.-June 1964,   HF240
      Annual 1964+   B8A4

BULGARIA

SA   BULGARIA. Tsentralno Statistichesko Upravlenie.
     Statisticheski godishnik na Narodna Republika Bulgariia. Statistical yearbook of the People's Republic of Bulgaria. 1956-

1959-63+	HA1621
1962-63 with Eng. trans.	S8
HL 1962-63+	HA1621 S797
HL 1962 Eng. trans.	S7972

SA   BULGARIA. Tsentralno Statistichesko Upravlenie.
     Statisticheski spravochnik na NR Bulgariia. 1958-

1958, 1961, 1963+	HA1621 S83

SA   BULGARIA. Glavna Direktsiia na Statistikata.
     Annuaire statistique du royaume de Bulgarie. 1909-42.

1909-35, 1938-42	HA1621 A5

SA   BULGARIA. Tsentralno Statistichesko Upravlenie.
     Statistical manual of the People's Republic of Bulgaria.

SA   BULGARIA. Tsentralno Statistichesko Upravlenie.
     Blagnostroiznave na naselenite. 1962(?)-

SB   BULGARIA. Glavna Direktsiia na Statistikata.
     Statistika. Quarterly, 1940-49(?)
     No longer published.

SB   BULGARIA. Glavna Direktsiia na Statistikata.
     Mesechni izvestiia.... Bulletin mensuel de la Direction Générale de la Statistique. Godinal, année 1908-48.
     Information in Bulgarian and French.
     Title varies: Mesechni statisticheski izvestiia. Bulletin statistique mensuel.

v. 26: 1937-v. 30: 1941;	314.972
v. 34: 1947-v. 35: 1948.	B933m
wanting: several issues	

SB   BULGARIA. Tsentralno Statistichesko Upravlenie.
     Statisticheski izvestiia. Statistical news. Quarterly.

CA   BULGARIA. Tsentralno Statistichesko Upravlenie.
     Vutroshna turgoviia na Narodna Republika Bulgariia; statistichoskii sbornik. The foreign trade of the People's Republic of Bulgaria; statistical handbook.

CA   BULGARIA. Direction Générale de la Statistique Rep. Bulgare.
     Statistique du commerce extérieur.

CA   BULGARIA. Glavna Direktsiia na Statistikata.
     Statistique du commerce ... avec les pays étrangers. 1880-1922(?)
     Text and tables in Bulgarian and French.
     Title varies slightly.

1913-15, 1916-20 (each issued in 1 vol.)	382.4972 B933
HL 1910-12, 1916-20	

CA **BULGARIA. Glavna Direktsiia na Statistikata.
     Vunshna turgoviia.... Commerce extérieur.... 1897-1903. 1, 1905//

CA **BULGARIA. Glavna Direktsiia na
      Statistikata.
      Desetgodishna statistika na vunshnata
      turgoviia. Statistique décennale du com-
      merce extérieur de la Bulgarie.
      1886/95-1896/1905//

AA BULGARIA. Glavna Direktsiia na
      Statistikata.
      Statistique agricole.

    1904-06, 1908-32,   630.64972
      1938/39, 1942-44   B933s
    wanting: 1924

BURMA (Union)

SA BURMA (UNION).
      Economic survey of Burma. 1951(?)-

    1952-63+   HC437
    wanting: 1953   B8B8

    FRI 1952-63+
           wanting: 1953

SA BURMA (UNION). Central Statistical and
      Economics Dept.
      Statistical yearbook. 1961-

SA BURMA (UNION).
      Burma yearbook and directory.
    1957/58-

SA BURMA (UNION). Central Statistical and
      Economics Dept.
      Statistical papers.

SA BURMA (UNION). Central Statistical and
      Economics Dept.
      Bulletin of statistics. 1951-
      Subseries of its Statistical papers.
      Includes annual statistics from 1938/39
      and monthly data from 1949-50.

    1951-60+   315.92
    wanting: 2d q. 1954,   B961
      1st q. 1957

    FRI Current two years
           only.

CA BURMA (UNION). Central Statistical and
CB   Economics Dept.
      Bulletin of export trade. Monthly.
      Dec. issue includes annual statistics.
      Issued June 1952-Mar. 1954 by the
      Office of the Collector of Customs and
      the Central Statistical and Economics
      Dept.

    1952-62+   382.592
    wanting: Mar., June 1960   B961e
                                f

CA BURMA (UNION). Central Statistical and
CB   Economics Dept.
      Bulletin of import trade. [Erratic,
      monthly and quarterly]
      Dec. issue includes annual statistics.

    June 1952-Dec. 1952   382.592
      (monthly); Feb. 1953-54   B961i
      (quarterly); 1955-61
      (monthly); 1962 (quar-
      terly)
    wanting: Nov. 1958, July
      and Aug. 1959, June
      1960

CA BURMA (UNION). Office of the Collector
      of Customs.
      Statement of the trade and navigation
    of Burma. 1946/47-1948/49//
      Continues Customs Dept. Statement of
    the sea-borne trade and navigation of
    Burma with foreign countries and Indian
    ports. 1885/86-1945/46.

    1946/47-1948/49   382.592
                                B963
                                f

CA BURMA (UNION). Customs Dept.
      Statement of the sea-borne trade and
    navigation of Burma with foreign coun-
    tries and Indian ports. 1885/86-1945/46.
      Continued by Office of the Collector of
    Customs. Statement of the trade and
    navigation of Burma. 1946/47-1948/49.

    1929/30-1930/31,   382.592
      1936/37, 1945/46   B962
                                f

AB BURMA (UNION). Dept. of Agriculture.
      Agriculture marketing bulletin.

    FRI July 1957, July 1958

AR BURMA (UNION). Dept. of Land Records and Agriculture.
   Report.

   1926   336.106
                              B962

BURUNDI

SB BURUNDI. Ministère des Affaires Economiques. Service de l'Information.
   Bulletin économique et financier.
   Monthly.

C

CAMBODIA

see also INDOCHINA

SA CAMBODIA. Direction de la Statistique et des Etudes Economiques.
   Annuaire statistique de Cambodge.
   1937-57--
   Title varies: 1937-57--1958-61, Annuaire statistique rétrospectif du Cambodge.

   1937-57--1962+   HA1751
                              A2

   FRI Current only.

SA INDOCHINA.
   Annuaire général de l'Indochine française. 1887-
   Continues Annuaire de l'Annam, Annuaire de la Cochinchine, Annuaire de l'Indochine, Annuaire du Cambodge.
   1890-97, pt. 1. Cochinchine et Cambodge; 1887-97, pt. 2. Annam et Tonkin.

   1924-25   354.596
                              A615

SA CAMBODIA.
   Annuaire du Cambodge. 1887-97//
   Continued in Indochina. Annuaire général de l'Indochine française.

SB CAMBODIA. Direction de la Statistique et des Etudes Economiques.
   Bulletin mensuel de statistique.

   1952-62+   HA1751
   wanting: several issues   A32
   1952-61   f

SB CAMBODIA. Direction de la Statistique et des Etudes Economiques.
   Bulletin économique.
   Vols. for 1951-52 issued by Direction de la Statistique, with title: Bulletin économique et statistique du Cambodge.

   1951-56   HA1751
   wanting: several issues   A3
                              f

CA CAMBODIA. Direction des Douanes et Régies.
   Bulletin de statistiques des échanges commerciaux, bulletin annuel.
   Limited number published, not generally available.

CAMEROONS (British (Administration)

   British Administration ended in 1961. Northern possessions joined Federation of Nigeria, Southern joined Republic of Cameroun.

SA GREAT BRITAIN. Colonial Office.
   Report by His Majesty's Government in the United Kingdom of Great Britain and Northern Ireland to the General Assembly of the United Nations on the administration of the Cameroons under the United Kingdom trusteeship.
   Issued in its series, Colonial no. -
   Has title: Report by His Britannic Majesty's Government to the Council of the League of Nations on the administration of the British Cameroons, 1925-38.
   1939-46 none published.

   1920/21   Brit. Docs.
                              1922
                              v. 16

   1922   916.694
                              G786

   1923-58+   325.342
                              G784
                              no. 6

CAMEROUN

French Administration ended 1960. Became Republic of Cameroun.

Oct. 1961- Federal Republic of Cameroun as joined by Southern Cameroons (formerly British Administration).

SA   CAMEROUN. Service de la Statistique.
Résumé des statistiques du Cameroun occidental. West Cameroun digest of statistics.

    1955/62 (1 vol.)   HA2117
                                 C3A4

SA   FRANCE. Service Colonial des Statistiques.
Annuaire statistique du Cameroun. Began publication in 1947 with vol. for 1938-45, 1947-57/58//

    1938-45   316.711
                                 F815

SA   FRANCE.
Rapport aux Nations Unies sur l'administration du Cameroun placé sous la tutelle de la France.

    1921-38, 1947, 1950-52   325.344
                                      F811
                                      f

SB   CAMEROUN. Service de la Statistique Générale et de la Mécanographie.
Résumé des statistiques du Cameroun oriental. Bulletin. Jan. 1963-

    Jan. 1963-Oct. 1965+   316.711
    wanting: Apr. 1963   C181r

SB   CAMEROUN. Service de la Statistique Générale et de la Mécanographie.
Note trimestrielle économique [sur la situation économique de la Fédération]. no. 1, 1964-

SB   CAMEROUN. Service de la Statistique Générale.
Bulletin de la statistique générale. v. 1, no. 1950-v. 12, June 1962//

    v. 6, no. 1-v. 12, no. 6;   316.711
    Oct./Dec. 1955-June   C181
    1962
    wanting: v. 7: 1, 2, 10;
      v. 8: 1-3, 9-10

SB   CAMEROUN. Service de la Statistique Générale.
Supplément au Bulletin de la Statistique Générale. Irregular, 1958-60. Discontinued.

CA   CAMEROUN. Service de la Statistique Générale.
Statistiques du commerce extérieur du Cameroun. 1946-

    v. 1, 1946/49-v. 8, 1961+   HF3914
                                       C3A5

    FRI Current two years
        only.

CA   CAMEROUN. Service de la Statistique
CB   Générale et de la Mécanographie.
Statistiques douanières, Cameroun oriental. 1960-

    1962, 1963, 1965 1st tri-   336.26
      mestre+   C181
                                       f

CA   CAMEROUN. Service des Douanes.
Le commerce extérieur du territoire du Cameroun. 1943-

    1953-61+   382.6711
                                   C182
                                   f

CANADA

SA   CANADA. Bureau of Statistics.
The Canada year book ... the official statistical annual of the resources, history, institutions, and social and economic conditions of the dominion.... 1905-

    1905-65+   HA744
                                S81

SA   CANADA. Dept. of Agriculture.
       The statistical year-book of Canada
.... 1886-1904.
       Continued by The Canada year book
(Canada. Bureau of Statistics).

1889-1904	317.1
wanting: 1900	C212c

SB   CANADA. Bureau of Statistics.
       Canadian statistical review. 1, Jan.
1926- Monthly.
       Beginning 1948, weekly supplement issued.
       Title varies.

1926-34, 1950-Mar. 1966+	330.971
wanting: v. 1, no. 1-6	C211
Weekly suppl. kept current year only.	

CA   CANADA. Bureau of Statistics.
       Trade of Canada. Commerce du Canada. 1930/31-
       1945- in 3 vols.: v. 1, Summary and analytical tables; v. 2, Exports; v. 3, Imports.
       1924/25-1929/30 issued in: Canada. Annual departmental reports (354.71 C212).

1931/32--1961-62 v. 2+	382.71
wanting: v. 1, 1940, 1941, 1944	C213t

CA   CANADA. Bureau of Statistics.
       Review of foreign trade.
       Suspended indefinitely after issue for 1960.

1947-54, 1960	382.71
	C2

CA   CANADA. Bureau of Statistics.
       Trade of Canada (imports for consumption and exports). 1916/17-

1927-38	382.71
	C212tr

CB   CANADA. Bureau of Statistics.
       Trade of Canada: summary of foreign trade. Monthly.

1947-Oct. 1965+	382.71
	C213tf

CB   CANADA. Bureau of Statistics.
       Trade of Canada: summary of exports. 1935-
       1935-46 has title Trade of Canada: domestic exports.
       Supersedes Exports of Canadian produce.

1935-46	382.71
	C213e
1947-Dec. 1965+	382.71
wanting: Jan. 1958, Aug. 1962	C213te

CB   CANADA. Bureau of Statistics.
       Trade of Canada: summary of imports. 1947-

1947-Oct. 1965+	382.71
	C213ti

CB   CANADA. Bureau of Statistics.
       Imports into Canada for consumption, by countries (excluding gold). (?) -Dec. 1946//
       Continued in its Trade of Canada: summary of imports.

Feb. 1935-Apr. 1942; Jan. 1944-Dec. 1946	382.71 C213i

CB   CANADA. Bureau of Statistics.
       Trade of Canada: exports by countries. Quarterly.

CB   CANADA. Bureau of Statistics.
       Trade of Canada: imports by countries. Quarterly.

CB   CANADA. Bureau of Statistics.
       Trade of Canada: exports by commodities.

CB   CANADA. Bureau of Statistics.
       Trade of Canada: imports by commodities.

CB   CANADA. Bureau of Statistics.
     ... Report of the trade of Canada
     (imports for consumption and exports).
     July 1894-
        Monthly, later quarterly.

     Sept. 1922-Apr. 1940   382.71
                                   C212tra

CB   CANADA. Bureau of Statistics.
        Census and statistics monthly.
        Superseded by Quarterly bulletin of
     agricultural statistics.

     FRI   v. 3-10, 1910-17

AA   CANADA. Bureau of Statistics.
        Agricultural statistics.

     FRI   1917, 1922-24,
              1925-29

AB   CANADA. Bureau of Statistics.
        Quarterly bulletin of agricultural sta-
     tistics.
        Supersedes its Monthly bulletin of
     agricultural statistics and Census and
     statistics monthly.

     FRI   1918-65+

AR   CANADA. Dept. of Agriculture.
        Report of the minister of agriculture.

     1917-19, 1924/25-1944/45,   630.671
        1947/48-1964/65+   C211r
     wanting: 1950/51

CANARY ISLANDS

SA   GREAT BRITAIN. Dept. of Overseas
CA   Trade.
        ... Report on the trade and economic
     conditions of the Canary Islands ....
     [1921-31.]

     1921, 1927   380.005
                                   G786

CAPE VERDE ISLANDS

see also PORTUGAL

SA   CAPE VERDE ISLANDS. Secção de
        Estatística.
        Anuário estatístico. 1933-

SA **CAPE VERDE ISLANDS.
        Estatística geral da Provincia de Cabo
     Verde .... 1912-(?)

SB   CAPE VERDE ISLANDS. Serviços de
        Administração. Secção de Estatística.
        Boletim trimestral de estatística.
     1949-

CA   CAPE VERDE ISLANDS. Secção de
        Estatística Geral.
        Estatística do comércio externo.
     1949-

CAYMAN ISLANDS

SA   GREAT BRITAIN. Colonial Office.
        Annual report on Cayman Islands ....
     1946-
        Previously issued in its numbered
     series, Colonial reports--annual
     (325.342 G787) no. 1092: 1918-19--
     no. 1872: 1937. Suspended 1940-46.

     1946--1959-60+   325.342
                                   G7871Ca

SA **CAYMAN ISLANDS.
        Blue book. 1913-(?)
        Includes statistics for all governmen-
     tal activities.

SB   CAYMAN ISLANDS.
        No statistical bulletin available; see
     Jamaica, Quarterly digest of statistics.

CENTRAL AFRICAN REPUBLIC

SA   CENTRAL AFRICAN REPUBLIC. Service de la Statistique Générale.
      Annuaire statistique de l'Oubangui. 1954-

SB   CENTRAL AFRICAN REPUBLIC. Service de la Statistique Générale.
      Bulletin mensuel de statistique.
1952 (?)
      Vols. -Nov. 1958 issued by Ubangi-Shari. Bureau de Statistique (called Bureau de la Statistique Générale, 1956-Nov. 1958).
      Title varies: 19(?)-56, Bulletin d'informations statistiques.

no. 41, May 1955-no. 59, Nov. 1956; no. 61, Jan. 1957-no. 161, May 1965+	HA2117 C4A3

SB   Union Douanière Equatoriale.
      Bulletin des statistiques générales.
no. 1, Jan. 1963-
      At head of title: Conférence des chefs d'état de l'Afrique équatoriale.
      Supersedes Bulletin mensuel de statistique formerly issued by Service de la Statistique Générale, French Equatoriale Africa. Quarterly.

no. 1, 1963-no. 12, 1965+	HF268 E7A52

SB   CENTRAL AFRICAN REPUBLIC. Service de la Statistique Générale.
      Bulletin trimestriel de statistique.
no. 1, 1955-no. 19, 1959//
      Vols. for 1955-58 issued by Ubangi-Shari, Bureau de la Statistique Générale.
      Title varies.

no. 1, 1955-no. 19, 1959	HA2117 C4A32

CA   Union Douanière Equatoriale.
      Commerce extérieur, U.D.E. et républiques composantes: Centrafrique, Congo, Gabon, Tchad.
      At head of title: Conférence des chefs d'état de l'Afrique équatoriale. Union Douanière Equatoriale. Statistiques générales.
      Supersedes Statistiques du commerce extérieur formerly issued by Service de la Statistique Générale, French Equatorial Africa.

1960-63	HF268 E7A5

CEYLON

SA   CEYLON. Dept. of Census and Statistics.
      Statistical abstract. 1949-
      Supersedes The Ceylon blue book.

1949-64+	HA1728 C43

SA   CEYLON. Dept. of Census and Statistics.
      The Ceylon year book.
      Began publication in 1948 covering period 1939-46.
      Supersedes Annual general report on Ceylon.

1948-62+	315.48 C427

SA   CEYLON. Dept. of Census and Statistics.
      The Ceylon blue book. 1821-1939.
      Superseded by its Statistical abstract.

1914-18, 1920-23, 1925	354.548 C429

SB   CEYLON. Dept. of Census and Statistics.
      Quarterly bulletin of statistics.
v. 1-11, Mar. 1950-Dec. 1960.
      Suspended indefinitely after Dec. 1960 issue.

Mar. 1950-Dec. 1960	315.48 C426q

CA   CEYLON. Dept. of Commerce.
      Administration report of the director of commerce. 1947/48-

1950-59, 1961/62-1963/64+ Report for 1960 NYP	382.548 C423

CA	CEYLON. Customs Dept. <u>Administrative report of the principal collector of customs.</u> 1898- Title varies: -1915, <u>Ceylon customs. Report</u>; 1916-24, <u>Customs and shipping. Report</u>; 1925-56, <u>Administration report on the customs and shipping.</u>				

CHAD

see also FRENCH EQUATORIAL AFRICA

CA   CEYLON. Customs Dept.
     <u>Administrative report of the principal collector of customs.</u> 1898-
     Title varies: -1915, <u>Ceylon customs. Report</u>; 1916-24, <u>Customs and shipping. Report</u>; 1925-56, <u>Administration report on the customs and shipping.</u>

1950-1962/63, pt. 1-2   382.548
                                       C424

SA   CHAD. Service de l'Information.
     <u>Annuaire de Tchad.</u>

1950/51   316.72
                A615

CA   CEYLON. Dept. of Import and Export Control.
     <u>Administration report of the controller of imports and exports.</u> 1950-

1950-1961/62+   382.548
                       C422

SB   CHAD. Service de la Statistique Générale.
     <u>Bulletin mensuel de statistique.</u>
1951(?)-
     Title varies: 195(?)- <u>Bulletin statistique.</u>
     Vols. for Nov. 1956- issued by French Equatorial Africa, Bureau de la Statistique du Tchad.

June 1956-Dec. 1965+   HA2117
wanting: issues 1958,   C5A3
  1959, 1961

CB   CEYLON. Customs Dept.
     <u>Ceylon customs returns.</u> 1904-
Monthly.

Dec. 1950-June 1965+   336.26
                                C425

SB   Union Douanière Equatoriale.
     <u>Bulletin des statistiques générales.</u>
no. 1, Jan. 1963-
     At head of title: Conférence des chefs d'état de l'Afrique équatoriale.
     Supersedes <u>Bulletin mensuel de statistique</u> formerly issued by Service de la Statistique Générale, French Equatorial Africa.

CB   CEYLON. Dept. of Commerce.
     <u>Ceylon trade journal.</u> Oct. 1935-

1951-May 1965+   382.548
                   C425

1: 1963-12: 1965+   HF268
                            E7A52

AA **CEYLON. Dept. of Agriculture.
     <u>Yearbook.</u> 1923-(?)

CA   CHAD. Service de la Statistique Générale.
     <u>Commerce extérieur.</u> 1961-62-63--
     Importations-exportations. Produits résumés ex-UDE.

AB   CEYLON. Dept. of Agriculture.
     <u>Bulletin.</u> 1912-(?)

no. 8-11, 18-19, 22,   630.6548
24-31, 35, 45-47,   C425
67-68: [1913-24]

1964-   HF268
         C5A3

AR   CEYLON. Dept. of Agriculture.
     <u>Administration report of the director of agriculture.</u>

1949-1962/63+   630.6548
                   C425a

CA   Union Douanière Equatoriale.
        Commerce extérieur, U.D.E. et républiques composantes: Centrafrique, Congo, Gabon, Tchad.
        At head of title: Conférence des chefs d'état de l'Afrique équatoriale. Union Douanière Equatoriale. Statistiques générales.
        Supersedes Statistiques du commerce extérieur formerly issued by Service de la Statistique Générale, French Equatorial Africa.

1960-63	HF268
wanting: 1961	E7A5

CHILE

SA GREAT BRITAIN. Dept. of Overseas
CA   Trade.
        ... Economic conditions in Chile ... Report .... [1921-37.]
        Title varies.

1921, 1923-25, 1927, 1929-30, 1932, 1934, 1936-37	380.005 G786

SA   CHILE. Dirección General de Estadística.
        Estadística anual. 1928-
        Continues Anuario estadístico, in 8 parts. Pt. 8 has title Anuario estadístico which contains Résumé of sections 1-7. See also CA and AA.

SA   CHILE. Dirección General de Estadística.
        Anuario estadístico. 1848/58-
        1848/58-1925, and several vols. for 1926, issued by Oficina Central de Estadística.
        Vols. for 1911- issued in subseries: Agricultura, etc., and are catalogued by subseries. See CA and AA.

1909: v. 1-3; 1910: v. 1-3	318.3 C537

SA **CHILE. Dirección General de Estadística.
        Statistical abstract of the Republic of Chile. 1916-(?)

SB   CHILE. Dirección de Estadística y Censos.
        Síntesis estadística.

Oct. 1962-Dec. 1965+	HA993 A5

SB   CHILE. Dirección de Estadística y Censos.
        Boletín. 1928-
        Title varies: 1928-Jan./Feb. 1961, Estadística chilena.

1928-36, Dec. 1946, 1949-Dec. 1962+	318.3 C537e
wanting: several issues	

SB **CHILE. Dirección General de Estadística.
        Monthly statistical report. v. 1-9: no. 8, 1919-27//

SB **CHILE. Dirección General de Estadística.
        Boletín estadístico. 1-9, May 1919-Oct. 1927//
        Superseded by Estadística chilena.

CA   CHILE. Dirección General de Estadística.
        Comercio exterior. 1916-
        1916-32 issued in its Anuario estadístico (1928-33 as Estadística anual); series numbering varies: 1916-27, v. 11; 1928-32, v. 7.
        Issuing agency varies.

1918-22, 1924-26, 1932, 1950-63+	382.83 C537a

CA   CHILE. Oficina Central de Estadística.
        Estadística comercial de la república de Chile. 1844-1915//
        Superseded by Chile. Dirección General de Estadística. Comercio exterior.

1909-11, 1915	382.83 C537

CA **CHILE. Oficina de Estadística Comercial.
Resúmenes estadísticos del comercio esterior de Chile ... Importación y esportación. 1873-1915//

AA CHILE. Dirección de Estadística y Censos.
Agricultura e industrias agropecuarias. 1911/12-
Vols. for 1911-27 issued as its Anuario estadístico; series numbering varies.
Supersedes Oficina Central de Estadística. Estadística agrícola.
Title varies: 1911/12-1930/31, Agricultura; -1931/32-1933/34, Estadística anual de agricultura.

1914/15-1925/26, 1943/44,   630.683
1948/49-1953/54,   C537a
1956--1961/62-1962/63

AA CHILE. Dirección General de Estadística.
Agricultura. 1928-
Pt. 3 of Estadística anual.

AA CHILE. Oficina Central de Estadística.
Estadística agrícola. 1877-78--
Superseded by Dirección de Estadística y Censos. Agricultura e industrias agropecuarias.

1909/10-1910/11   630.683
                               C537

AB CHILE. Ministerio de Agricultura.
Agricultura y ganadería. Quarterly.

FRI no. 6, 7-27; 1956,
     1958-62, 1965+

AR CHILE. Ministerio de Agricultura e Industria.
Memoria de los ministerios de agricultura e industria y de tierras y colonización. 1926-

1926   630.683
                               C536

CHINA

SA GREAT BRITAIN. Dept. of Overseas
CA   Trade.
... Report on the industrial and economic situation of China ... . [1919-37.]
1919 issued in the series of Parliamentary Papers as Papers by command. Cmd. 853.
1923 (June) includes also a Report on the trade of Chinese Turkestan.
1928, 1930 include Report on the trade of South Manchuria; 1935-37 includes an annex on economic conditions in Hongkong.
Title varies.

1919, 1921-37   380.005
                               G786

HL   1928-30, 1935-37   HC426
                               G786

SB CHINA. Bureau of Statistics.
The statistical monthly.
June 1935-Dec. 1936 published as Quarterly journal of statistics.

no. 9-30; 1933-37   315.1
                               S797

CA CHINA. Inspectorate General of Customs.
... Foreign trade of China. 1920-31//
In two parts each year, 1920-31: pt. 1, Report and abstract of statistics; pt. 2, Analysis--v. 1, Imports; v. 2, Exports.
Continued by The trade of China. 1933-

1920-31   382.51
                               C539r

CA CHINA. Inspectorate General of Customs.
Annual trade report and returns. 1920-31.
Quarterly for each port issued separately but included with this series.

1920-28, [1929-31]   382.51
                               C539a

CA CHINA. Inspectorate General of Customs.
... Decennial reports on the trade, navigation, industries, etc. ... .
1822/91-(?)

1902-11, 1912/21,   382.51
   1922/31   C539d

CA   CHINA. Inspectorate General of Customs.
     Returns of trade ... and trade reports.
1882-1919//
     Continued in part by its Foreign trade of China and Annual trade report and returns, 1920-31.

    1898: no. 6, 1904,   382.51
      1908-19   C539r
    Each year represented
      but not all parts
      each year.

CA **CHINA. Inspectorate General of Customs.
     Reports on trade at the treaty ports.
1-17, 1865-81//
     13-17, 1877-81 published as Maritime customs, pt. 1, no. 4. Continued as a part of Returns of trade and trade reports.

CA **CHINA. Inspectorate General of Customs.
     Returns of trade at the treaty ports.
1-23, 1859-81//
     Issued in two parts each year.
     19-23, 1877-81, published as Maritime customs, pt. 1, no. 3.
     Continued as a part of Returns of trade and trade reports.

CB   CHINA. Inspectorate General of Customs.
     Monthly returns of the foreign trade of China. Nov. 1931-
     Issued as its Chinese maritime customs. I. Statistical series no. 8.
     Publication suspended May 1943-June 1946.

    Jan. 1932-Dec. 1948   HF237
    wanting: several issues   A44

CB   CHINA. Inspectorate General of Customs.
     Quarterly returns of trade. no. 1-204, 1869-1919.

    no. 161-204, 1909-19   382.51
                                   C539c

AA **CHINA. Ministry of Agriculture and Commerce (Nung-shang Pu).
     Statistical report. (Nung-shang tung chi piao.) 1-9, 1912-20//

CHINA (People's Republic of China, 1949-)

SA   CHINA (PEOPLE'S REPUBLIC OF CHINA, 1949-). State Statistical Bureau.
     Ten great years; statistics of the economic and cultural achievements of the People's Republic of China. 1960.

    1960   330.951
                                  C537b

HL   1960   HC427.1
                                  A362
HL   Original in Chinese   A362a

CHINA (Taiwan)

SA   CHINA (TAIWAN). Directorate-General of Budgets, Accounts and Statistics.
     Statistical abstract of the Republic of China.

    1947, 1959-65+   HA1706
                                   C4
    FRI   1947

SA   CHINA (TAIWAN). Council for International Economic Cooperation and Development.
     Taiwan statistical data book ....

    1963-65+   HA1706
                                  T3

CA   CHINA (TAIWAN). Inspectorate General of Customs.
     The trade of China.
     Supersedes Foreign trade of China, 1920-32, and Returns of trade and trade reports, 1904-20.

    1933-43, 1946-48,   382.51
      1950-64+   C539r
                                  1932-

AA   TAIWAN (PROVINCE). Dept. of Agriculture and Forestry.
     Taiwan agricultural yearbook. 1946-1948- with English translation included.

    FRI   1948-49, 1952-53,
           1961-65+

AA Joint Commission on Rural Reconstruction in China (U.S. and China).
    <u>Taiwan agricultural statistics</u>.
    1901-55 (1 vol.)
    Issued as its Economic digest series no. 8.

    HL   1901-55   HD2095
                                     F72J74

    FRI   1901-55

COCHIN-CHINA

SA **COCHIN-CHINA.
    <u>Annuaire de la Cochinchine</u>. 1865-1900/01//
    1865-71 as <u>Annuaire de la Cochinchine française</u>. Continued in: Indo-China. <u>Annuaire général de l'Indochine française</u>.

COLOMBIA

SA COLOMBIA. Departamento Administra-
CA   tivo Nacional de Estadística.
        <u>Anuario general de estadística</u>. 1905, 1915-
        Some vols. issued in parts.

        1918-22, 1923, 1930,   HA1011
        1935-36, 1938-61+   A16

        FRI   1946-62

SA GREAT BRITAIN. Dept. of Overseas
CA   Trade.
        ... <u>Republic of Colombia. Commercial review and handbook</u> .... [1921-38.]
        Title varies.

        1921, 1924, 1925,   380.005
        1929, 1938   G786

SB COLOMBIA. Departamento Administrativo Nacional de Estadística.
    <u>Boletín mensual de estadística</u>. 1951-
    Title varies: 1951-July 1952, <u>Boletín informativo</u>.

    Jan. 1952-Dec. 1964+   318.4
    wanting: Oct. 1952,   C7b
        Apr. 1958

SB COLOMBIA. Departamento de Contraloría.
    <u>Anales de economía y estadística</u>.
    v. 1, 1935-no. 86, 1959//
    Issued irregularly, mostly monthly.
    Data vary from issue to issue.
    1957(?)- has title <u>Economía y estadística</u>.

    v. 1-7: 1935-44; 2d epoca: 330.984
        no. 1-72, 1945-50; 3d   C7171
        epoca: no. 73-76, 1952;
        4th epoca: no. 81,
        83-84, 1957-58
    occasional supplements

SB COLOMBIA. Departamento de Contraloría.
    ... <u>Información fiscal de Colombia</u> .... 1927-40.
    Monthly; some months combined; none published for Jan. 1936.

    1932-40   318.4
    very incomplete   C7i

SB **COLOMBIA. Departamento de Contraloría.
    <u>Boletín de estadística</u>. 1-15, 1912-26//(?)

CA COLOMBIA. Departamento Administrativo Nacional de Estadística.
    <u>Anuario de comercio exterior</u>.
    Vols. for 1920, 1921, not issued.

    1916, 1918-19, 1923,   382.84
    1933-63+   C718

AA COLOMBIA. Departamento Administrativo Nacional de Estadística.
    <u>Directorio nacional de explotaciones agropecuarias</u>.

    FRI   1960, 1964

AA COLOMBIA. Ministerio de Agricultura.
AR   División de Economía Rural.
        <u>Economía agropecuaria de Colombia</u>.

        FRI   1948, 1950

AR   COLOMBIA. Departamento de Agricultura.
     ...Informe....

     1938/39   630.684
                                C717

CONGO, BELGIAN

SA   CONGO, BELGIAN. Gouverneur Général.
     Discours.   -1959//
     Vols. for 1953-57 include a separately paged section: Statistique.
     Statistics for 1957-59 published as Bulletin annuel des Statistiques du Congo belge; cover title: Statistiques relatives à l'année.

     1947-59   354.675
                                C749

SA   BELGIUM. Ministère des Affaires Africaines.
     Rapport sur l'administration de la colonie du Congo belge.   -1958//

     1925, 1927, 1929,   354.675
        1932-36, 1938-58   B429r
     wanting: 1948   f

SB   CONGO, BELGIAN. Direction de la Statistique.
     Bulletin mensuel des statistiques générales du Congo belge et du Ruanda-Urundi. v. 1-6, 1955-61//
     Superseded by Congo (Leopoldville). Direction de la Statistique. Bulletin des statistiques générales.
     Contains data formerly in the agency's Bulletin mensuel des statistiques du Congo belge et du Ruanda-Urundi, later Bulletin mensuel de commerce extérieur. See Congo (Leopoldville) CB.

     v. 1-5; 1955-60   316.75
     v. 5 has only no. 1, 2, 4 C749bg

CA   CONGO, BELGIAN. Secrétariat Général.
     Statistique du commerce extérieur de l'union douanière du Congo belge et du Ruanda-Urundi. 1941-58//
     Superseded by Congo (Leopoldville). Bulletin du commerce extérieur, 1958-63, and Annuaire commerce extérieur, 1964-

     1914-1948/49, 1951-58   382.675
                                C749

AA   BELGIUM. Services de l'Agriculture du Congo Belge.
     L'Agriculture au Congo belge et au Ruanda-Urundi.

CONGO (Brazzaville)

SA   CONGO (BRAZZAVILLE). Service National de la Statistique, des Etudes Démographiques et Economiques.
     Annuaire statistique.

SB   CONGO (BRAZZAVILLE). Service National de la Statistique, des Etudes Démographiques et Economiques.
     Bulletin mensuel rapide de statistique.
     no. 1, Jan./May 1963-

     Jan./May 1963-   HA2031
     Nov. 1965+   A7

SB   Union Douanière Equatoriale.
     Bulletin des statistiques générales.
     no. 1, Jan. 1963-
     Supersedes Bulletin mensuel de statistique formerly issued by Service de la Statistique Générale, French Equatorial Africa.

     no. 1, 1963-   HF268
     no. 12, 1965+   E7A52

SB   CONGO (BRAZZAVILLE). Service de la Statistique.
     Bulletin mensuel de statistique. 1957-

     1960: no. 40, 41, 45   316.75
     1961: no. 5, 6   C749bs
     1962: no. 01, 02, 3

CA   Union Douanière Equatoriale.
     Commerce extérieur, U.D.E. et républiques composantes: Centrafrique, Congo, Gabon, Tchad.
     Supersedes Statistiques du commerce extérieur formerly issued by Service de la Statistique Générale, French Equatorial Africa.

     1960-63+   HF268
                                E7A5

CA CONGO (BRAZZAVILLE). Service de la
     Statistique.
     Estimation du commerce extérieur du
     Congo en 1959. (1 vol.)
     Cover title: Etude des importations et
     des exportations de la République du
     Congo en 1959.

     1959   382.6424
                                       C747g

AR CONGO (BRAZZAVILLE). Service de
     l'Agriculture.
     Rapport pour l'annee.

CONGO (Leopoldville)

SB CONGO (LEOPOLDVILLE). Direction de
     la Statistique et des Etudes Econo-
     miques.
     Bulletin des statistiques générales.

     v. 7, no. 1-v. 11, no. 3;   HA2031
     Jan. 1962-Sept. 1965+   A4

SB CONGO (LEOPOLDVILLE).
     Bulletin accéléré. v. 1, 1963-

CA CONGO (LEOPOLDVILLE). Direction de
     la Statistique.
     Annuaire des statistiques du com-
     merce extérieur. 1964-
     Supersedes its Bulletin du commerce
     extérieur de la Republique du Congo.

CB CONGO (LEOPOLDVILLE). Direction de
     la Statistique.
     Bulletin du commerce extérieur de la
     République du Congo. v. 1, 1950-
     v. 1-5 also called no. 1-50.
     Title varies: 1950-54, Bulletin men-
     suel des statistiques du Congo belge et du
     Ruanda-Urundi; 1955-60, Bulletin men-
     suel du commerce extérieur du Congo
     belge et du Ruanda-Urundi.
     Superseded by Annuaire des statis-
     tiques du commerce extérieur. 1964-

     v. 3, no. 26; v. 4, no. 32;   382.675
     v. 10, no. 12; v. 11-14+   C749b
     [1952-59], 1960-63

CB CONGO (LEOPOLDVILLE). Direction de
     la Statistique et des Etudes Econo-
     miques.
     Etats mécanographiques exportations
     (exports et réexports par chapitres, pays
     et produits-pays). 1964-

CB CONGO (LEOPOLDVILLE). Direction de
     la Statistique et des Etudes Econo-
     miques.
     Etats mécanographiques importations
     (par chapitres, pays et produits-pays).
     1964-

COOK ISLANDS

SA NEW ZEALAND. Dept. of Island Terri-
     tories.
     Reports on the Cook, Niue, and
     Tokelau Islands.
     Supersedes the Dept.'s Cook Islands
     annual report and its Tokelau Islands
     annual report.

     1947/48-1949/50,   354.9623
     1951/52-1964/65+   N532

COSTA RICA

SA COSTA RICA. Dirección General de
     Estadística.
     Anuario estadístico. 1883-93, 1907-45,
     1948-

     1911-16, 1918-19,   HA802
     1921-22, 1925-31,   A2
     1934-45, 1948-64+

SA COSTA RICA. Dirección General de
     Estadística.
     Informe. 1897-1908, 1920-47//

     1926-28, 1939, 1944-47   317.286
                                   C837i

SA GREAT BRITAIN. Dept. of Overseas
CA   Trade.
        ... Economic conditions in the republic of Costa Rica ... Report ....
        [1925-33.]
        Statistics for 1921-22, 1926, 1929, and 1933/35- are found in the Report on economic and commercial conditions in the republic of Panama and the Panama canal zone ... and in the republic of Costa Rica ... (Gt. Brit. Dept. of Overseas Trade).
        Title varies.

        1925, 1929-33   380.005
                                     G786

SB COSTA RICA. Dirección General de
        Estadística y Censos.
        Boletín. Series estadísticas. July 1950-
        Discontinued.

CA COSTA RICA. Dirección General de
        Estadística y Censos.
        Comercio exterior de Costa Rica.

        1953-64+   382.7286
                                     C837co

CA COSTA RICA. Dirección General de
        Estadística y Censos.
        Estadísticas de exportación.

        1949   382.7286
                                     C837ex

CA COSTA RICA. Dirección General de
        Estadística y Censos.
        Estadísticas de importación.

        1949   382.7286
                                     C837ei

CA COSTA RICA. Dirección General de
        Estadística y Censos.
        Estadística de comercio exterior; importación y exportación por artículos.
        1886-1950(?)

        1946-49   382.7286
                                     C837e

CA COSTA RICA. Dirección General de
        Estadística y Censos.
        Estadísticas de comercio exterior. Resumen. (?) -1950//

        1947-48   382.7286
                                     C837ee

CA COSTA RICA. Dirección General de
        Estadística y Censos.
        Boletín de exportación. no. 1, 1941-

        1944-46   HF137
                                     A18

CA COSTA RICA. Dirección General de
AA   Estadística y Censos.
        Resúmenes estadísticos, años 1883 a 1910. Comercio, agricultura, industria.

        1883/1910   317.286
                                     C837r

CB COSTA RICA. Dirección General de
        Estadística y Censos.
        Boletín de comercio exterior. no. 1, Aug. 1953-
        Discontinued.

AA COSTA RICA. Ministerio de Agricultura e Industrias.
        Memoria.
        Title varies: 19(?)--1960-61, Informe anual.

        1958, 1960-61, 1962-63+   S169
                                     A34

CUBA

SA CUBA. Dirección General de Estadística.
        Anuario estadístico de Cuba.
        Anuario estadístico de la República de Cuba, 1914, only issue published until 1952.

        1952, 1956   HA871
                                     A3

SA   CUBA.
       Resumen estadístico. Statistical summary [for Cuba].

SB   CUBA. Dirección General de Estadística.
       Boletín mensual de estadísticas.
   1945-55//
       Series erratic but complete.

     v. 1-11, 1945-June 1955   317.291
     wanting: v. 2: 1;   C964
       v. 4: 1-11

CA   CUBA. Dirección General de Estadística.
       Comercio exterior. 1902-1957/58.

     1910-1957/58   382.7291
     wanting: 1914-18, 1923,   C962
      1930-34, 1936-43

CA   CUBA. Dirección General de Estadística.
       Comercio exterior de Cuba. Exportación.
       Continues in part Comercio exterior.

CA   GREAT BRITAIN. Dept. of Overseas
SA   Trade.
       ... Economic conditions in Cuba ...
   Report .... [1922-37.]
       Title varies.

     1922, 1925, 1927, 1929,   380.005
      1932, 1935, 1937   G786

CA   CUBA. Dirección General de Estadística.
       Importación y exportación. 1923-

     1923-26   382.7291
                                 C962i

CB   CUBA. Ministerio del Comerio Exterior.
       Comercio exterior. v. 1, 1963-
       Supersedes a publication with the same title, formerly published by Dirección General de Estadística.

     June-Dec. 1963; Jan.-   HF149
      Sept. 1964; Jan.-June   A3
      1965--Oct.-Dec. 1965+

CB   CUBA. Dirección General de Estadística.
       Sección de Comercio.
       Exportación. Sept. 1950-
       Supersedes in part its Boletín.

     Sept.-Dec. 1950;   382.7291
      Jan./June 1951; Jan.-   C963e
      Aug., Oct.-Dec. 1952;
      Jan.-Dec. 1954; Jan.-
      June 1955; Jan.-Feb.
      1956

CB   CUBA. Dirección General de Estadística.
       Sección de Comercio.
       Importación. Sept. 1950-
       Supersedes in part its Boletín.

     July-Dec. 1950;   382.7291
      Jan./June 1951;   C963i
      Jan.-Dec. 1952;
      Jan. 1953

CB   CUBA. Dirección General de Estadística.
       Sección de Comercio.
       [Boletín].
       Ceased with issue for Aug. 1950.
       Superseded by its Importación and its Exportación.

     June-Aug. 1950   382.7291
                                   C963b

AA   CUBA. Secretaria de Agricultura, Industria y Trabajo.
       Estadística agro-pecuaria. 1928/29-(?)
       Name of ministry varies.

CURAÇAO

see NETHERLANDS ANTILLES

CYPRUS

SA   CYPRUS. Statistics and Research Dept.
       Statistical abstract. 1955-
       Issued 1955- by the Financial Secretary's Office, Statistics Section.

     1959-64+   HA1950
                                   C9A3

SA   GREAT BRITAIN. Colonial Office.
     Report on Cyprus. 1946-
     Previously issued in its numbered
series, Colonial reports--annual
(325.342 G787) no. 1093: 1920-
no. 1895: 1938. Suspended 1940-46.

     1946-59+   325.342
                                  G7871cy

SA   CYPRUS.
     The Cyprus blue book. 1878-1946(?)

     1914-19, 1946   354.564
                                  C996
                                  f

SA   GREAT BRITAIN. Dept. of Overseas
CA   Trade.
     ... Report on economic conditions in
Cyprus and Malta. 1935.

     1935   380.005
                                  G786

SB   CYPRUS. Statistics and Research Dept.
     Statistical summary for the month.
1, July 1950-

     Jan. 1964-Dec. 1965+   HA1950
                                  C9A4
                                  f

CA   CYPRUS. Statistics and Research Dept.
     Statistics of imports and exports.
1954-
     Supersedes in part Cyprus. Comp-
troller of Customs and Excise. Statistics
of imports, exports and shipping ....

     1954-55, 1963+   HF240
                                  C9A33

CA   CYPRUS. Comptroller of Customs and
        Excise.
     Statistics of imports, exports and
shipping 1926-53//
     Superseded by Cyprus. Statistics and
Research Dept. Shipping statistics and
Statistics of imports and exports.

     1945-53   382.564
                                  C964a

AR   CYPRUS. Agricultural Dept.
     Annual report. 1895-

     1920/21, 1923-25,   630.6564
       1946-63+   C996
     wanting: 1956, 1958

AB   CYPRUS. Agricultural Dept.
     Bulletin B. 1956-

AB   CYPRUS. Agricultural Dept.
     Six-monthly review. 1955-
     Supersedes its Quarterly review.

CZECHOSLOVAK REPUBLIC

During the earlier years of the Republic, official statistics appeared in several series and in several language editions. Each series appeared in subseries by subject and/or part number. The series are listed below; Hoover Library has catalogued this material according to their subseries.

Those subseries which fall within the scope of this bibliography are listed individually under designated codes (SA, CA, etc.); the others do not appear individually.

          CZECHOSLOVAK REPUBLIC. Státní
             Urad Statistický.
          Ceskoslovenská statistiká. 1922-48.

          HL v. 31, 37, 57, 60, 103,
                 117-120, 122-125,
                 127-130, 135-139,
                 141-146, 150-153,
                 157, 160-173, 175-
                 177, 179, 180

          CZECHOSLOVAK REPUBLIC. Státní
             Urad Statistický.
          La Statistique tchécoslovaque. v. 1-
          140, 1922-37.

          HL v. 1-140.
                 wanting: v. 77-78, 93,
                 98, 102-103, 117-120,
                 122-125, 127-130,
                 135-136, 138-139

CZECHOSLOVAK REPUBLIC. Státní
   Urad Statistiký.
   Cechoslovakische Statistik.

HL   Bd. 102, 161-165

SA   CZECHOSLOVAK REPUBLIC. Ustrední
        Komise Lidové Kontroly a Statistiký.
     Statistická rocenka Ceskoslovenské
Socialistické Republiký. 1934-
     Publication suspended 1939-56.
     Vols. for 1934-60 issued by Státní
Urad Statistiký; 1961-62 by Ustrední
Urad Státní Kontroly a Statistiký; 1963-
by Ustrední Komise Lidové Kontroly a
Statistiký.
     Continuation of Statistická prirucka,
issued by Státní Urad Statistiký.

     1957-64+   HA1191
                           A416

HL   1957-64+   HA1191
     wanting: 1959   A47

FRI Current two years
         only.

SA   CZECHOSLOVAK REPUBLIC. Státní
        Urad Statistiký.
     Czechoslovak statistical abstract.
v. 1, 1958-

HL   v. 1, 1958-   HA1191
                           A4162

CA   CZECHOSLOVAK REPUBLIC.
     Translation and glossary of Czecho-
slovak statistical yearbook, 1957-
issued by U.S. Joint Publications
Research Service.

     1959   JPRS 3104   Monthly Catalog
                           June 1960
                           8602

SA   CZECHOSLOVAK REPUBLIC. Státní
        Urad Statistiký.
     Statistical digest of the Czechoslovak
Republic. 1948-

     1948   314.37
                           C998s

HL   1948   HA1191
                           A52

FRI 1948

SA   CZECHOSLOVAK REPUBLIC. Státní
        Urad Statistiký.
     Annuaire statistique. v. 1-5, 1934-38.
     Supersedes its Manuel statistique de
la République.

HL   1934-38   HA1191
                           A44

SA   CZECHOSLOVAK REPUBLIC. Státní
        Urad Statistiký.
     Statistische Uebersicht für die
Cechoslovakische Republik.

HL   1930, 1936   HA1191
                           A4713

SA   CZECHOSLOVAK REPUBLIC. Státní
        Urad Statistiký.
     Aperçu statistique de la République
Tchécoslovaque.
     1930-   includes figures from
1921-29.

HL   1930   HA1191
                           A4714

SA   GREAT BRITAIN. Dept. of Overseas
CA   Trade.
     ... Report on economic and commer-
cial conditions in Czechoslovakia ....
1921-

     1921, 1923-25, 1927, 380.005
     1929, 1931, 1933,   G786
     1935, 1937

SA   CZECHOSLOVAK REPUBLIC. Státní
        Urad Statistiký.
     Manuel statistique de la République
....
     Etabli par l'Office de Statistique
d'Etat. 1920-32.
     Superseded by its Annuaire statistique,
1934-38.

HL   1920-32   HA1191
     1928 German ed.   A43

SA   CZECHOSLOVAK REPUBLIC. Ustrední
        Urad Státní Kontroly a Statistiký.
     Cisla pro kazdeho. Mala statistická
rocenka.

SB   CZECHOSLOVAK REPUBLIC. Ustrední
      Komise Lidové Kontroly a Statistiký.
      <u>Statistiká, ekonomicko--statistiký
      casopis.</u>

      HL no. 1-12, 1964;
           no. 1-12, 1965+
          supplements: no. 1,
           1965+

SB   CZECHOSLOVAK REPUBLIC. Ustrední
      Komise Lidové Kontroly a Statistiký.
      <u>Statistiká a kontrola.</u> (?) -1963//

      HL no. 1-12, 1962;
           no. 1-12, 1963

SB   CZECHOSLOVAK REPUBLIC. Státní
      Urad Statistický.
      <u>Statistický obzor: statistical review
of Czechoslovakia.</u> 1931-61.
         Title varies: 1920-31, <u>Statistiký
vestnik.</u>
         Superseded by <u>Statistiká a kontrola.</u>

      HL 1920-56, 1961
           wanting: several
             issues

SB   CZECHOSLOVAK REPUBLIC. Ustrední
      Komise Lidové Kontroly a Statistiký.
      <u>Statistickezpravy.</u> (?) -1963//

      HL 1956-63
           wanting: 1956: 3-4

SB   CZECHOSLOVAK REPUBLIC. Státní
      Urad Statistický.
      <u>Statistical bulletin of Czechoslovakia.</u>
      Ceased publication 1949.

      HL v. 2: no. 1-6, July-   HA1191
          Dec. 1947   A532
         v. 3: no. 1-10, Jan. -
          Dec. 1948

SB   CZECHOSLOVAK REPUBLIC. Státní
      Urad Statistický.
      <u>Statistický zpravodaj: Bulletin statis-
tique Tchécoslovaque.</u> Jan. 1938-
         1938-48: in Czech only; Jan. 1949-
in four languages. Supersedes three
separate language publications.

      HL 1938, 1946-50

CA   CZECHOSLOVAK REPUBLIC. Státní
      Urad Statistický.
      <u>Ceskoslovenská statistiká: Rada 3.
Zahranicni obchod.</u> 1939-45.

      HL sesit 45-47: v. 160-   HF192
           162, 164, 165,   C9A3
           167, 169, 171,
           173, 175, 177,
           179, 180

CA   CZECHOSLOVAK REPUBLIC. Státní
      Urad Statistický.
      <u>Cechoslovakische Statistik: Reihe 3.
Aussenhandel.</u>

      HL Heft 46-49; Bd. 161,   HF192
           162, 164, 165;   C9A33
           1940-41

CA   CZECHOSLOVAK REPUBLIC. Státní
      Urad Statistický.
      <u>La Statistique tchécoslovaque: 3.
série. Commerce extérieur.</u> livr. 1-36,
1922-36.

      HL v. 1-24, 26-36;   HF192
           v. [3-133]   C9A34
           1922-36

CA   CZECHOSLOVAK REPUBLIC. Státní
      Urad Statistický.
      <u>Zahranicni obchod Republiký
Ceskoslovenské.</u> Foreign trade.

      HL 1923-24

CB   CZECHOSLOVAK REPUBLIC. Státní
      Urad Statistický.
      <u>Misicní prehled zahranicniho obchodu
Republiký Ceskoslovenské ... Obchod
speciální vydal Státní Urad Statistický.</u>
1923-

      HL 1925-35, 1939,
           1946-48

CB   CZECHOSLOVAK REPUBLIC. Státní
      Urad Statistický.
      <u>Zahranicni obchod Republický
Ceskoslovenská. Obchod speciální.</u>
Foreign trade, special.

      HL no. 1-12: 1923;
           no. 1-12: 1924

CB CZECHOSLOVAK REPUBLIC. Státní
    Urad Statistický.
    Statistický prehled zahranicniho
    obchodu Republický Ceskoslovenské za
    rok. 1920-22.
    Superseded by its Misicní prehled
    zahranicniho obchodu.

    HL 1920: 2 vols.

CB CZECHOSLOVAK REPUBLIC. Státní
    Urad Statistický.
    Aperçu statistique du commerce
    extérieur de la République Tchécoslo-
    vaque .... Monthly.

AA CZECHOSLOVAK REPUBLIC. Státní
    Urad Statistický.
    Ceskoslovenská statistiká: Rada 12.
    Zemedelstvi.

    HL sesit 22-23; v. 143,   HD1940
        166; 1937-41   C9A4

AA CZECHOSLOVAK REPUBLIC. Státní
    Urad Statistický.
    Ceskoslovenská statistiká: Rada 16.
    Scitani zemedelskych zavodu.

    HL sesit 19; v. 135;   HD1940
        1936   C9A5

AA CZECHOSLOVAK REPUBLIC. Státní
    Urad Statistický.
    Cechoslovakische Statistik: Reihe 12.
    Landwirtschaft.

    HL Heft 21; Bd. 102;   HD1940
        1934   C9A43

AA CZECHOSLOVAK REPUBLIC. Státní
    Urad Statistický.
    La Statistique tchécoslovaque: 12.
    série. Agriculture. livr. 1-20, 1922-33.

    HL v. 1-20; 1922-33   HD1940
        wanting: livr. 8   C9A44

    FRI v. 1-7, 9, 12-16:
        1922-28

AB CZECHOSLOVAK REPUBLIC. Státní
    Urad Statistický.
    Vestnik. 1960-

AB CZECHOSLOVAK REPUBLIC. Minister-
    stvo Zemedelstvi.
    Vestnik ....

    HL 1919-26, 1935-38

AB CZECHOSLOVAK REPUBLIC. Minister-
    stvo Zemedelstvi.
    Publication du ministère de l'agricul-
    ture de la République Tchécoslovaque.
    1920-22 issued as its Bulletin.

    HL 1921-38   HD1940
        wanting: 4th quarter   C9A314
        1924

D

DAHOMEY

see also FRENCH WEST AFRICA before 1958

SB DAHOMEY. Service Central de la Statis-
    tique et de la Mécanographie.
    Bulletin de statistique.

    v. 2, no. 4-6; Apr.   HA2117
    1965-Sept. 1965+   D3A4

SB DAHOMEY. Direction des Affaires Eco-
    nomiques.
    Bulletin économique et statistique.
    no. 1, Jan. 1953-

    1955-65+   HC547
    wanting: 1955: 1-2; 1958:   D3A4
    1-4; 1959: 4-12; 1960:
    1-8; 1962: 2

CA DAHOMEY.
    Commerce spécial; importation et
    cabotage à l'importation. 1958(?)-

AR DAHOMEY. Service de Développement
    Rural.
    Rapport annuel. 1961-

    FRI 1963+

DENMARK

SA   DENMARK. Statistiske Departement.
       Statistisk aarbog.   Annuaire statistique.   1896-

     1910-65+   HA1477
     wanting: 1962   A2

SA   DENMARK. Statistiske Departement.
       Statistisk oversigt.
       "Tillaig til 'Statistiske efterretninger.'"

     1948-59, 1950-60,   314.89
        1951-61+   D397so

SA   DENMARK. Statistiske Departement.
       Danmarks statistik.   Statistique de Danemark.   Statistiske meddelelser .... Communication de statistique.   1897-1958.
     Statistics continued in individual series by subject issued by Statistiske Departement.
     Issued in 4 chronological series: ser. 1, v. 1-6: 1852-1861; ser. 2, v. 1-13: 1862-77; ser. 3, v. 1-18: 1879-97; ser. 4, v. 1-174: 1897-1958.

     ser. 3, v. 18, pt. 1;   314.89
       ser. 4, v. 29-174   D397s
     wanting: v. 72: 1; v. 73: 2;
       v. 76: 1-3; v. 78: 1-3;
       v. 79: 1; v. 80: 1.

SB   DENMARK. Statistiske Departement.
       Statistiske efterretninger.   v. 1, 1909-
     Irregular, weekly plus 1 each year.

     v. 1-57, 1909-65+   314.89
     wanting: several issues   D397se
       for v. 6-12

CA   DENMARK. Statistiske Departement.
       Danmarks vareindførsel og -udførsel. Foreign trade of Denmark.   1898-
     1898-1958 issued in its series Statistisk tabelvaerk, included as a numbered publication in this series (314.89   D397t Litre D).

     1959-63+   382.489
     1909-58 shelved with   D397d
       Statistik tabelvaerk

CB   DENMARK. Statistiske Departement.
       Vareomsaetningen med udlanted i aaret.   v. 1, 1910-
     Issued in series Danmarks statistik, handelsstatistisk meddelelser, bd. 1-

     bd. 1-56, 1910-65+   382.489
     wanting: issues for   D397v
       1941-Aug. 1945

AA   DENMARK. Statistiske Departement.
       Landbrugsstatistik herunder gartneri, skovbrug m. v.   Statistics on agriculture, gardening and forestry.   1936-
     Issued as a numbered (later unnumbered) publication in its series Danmarks statistik ....   See under SA for issues previous to 1959.

     FRI 1959-64+

AA   DENMARK. Statistiske Departement.
       Høsten i Danmark.   Récoltes en Danemark.   1897-1935//
     Issued in its series Danmarks statistik.   See reference under SA

DOMINICA

SA   GREAT BRITAIN. Colonial Office.
       Annual report on Dominica: B. W. I.   1947-
     Previously issued in its numbered series, Colonial reports--annual (325.342   G787) under Leeward Islands; no. 1074: 1919-20--no. 1934: 1938. Suspended 1940-46.

     1947-60+   325.342
                                 G7871Do

SA   **DOMINICA.
       Blue book.   1821-87//(?)

SB   DOMINICA.
     No statistical bulletin available. Data on cost of living, imports and exports appear in the Dominica official gazette.

CA	DOMINICA. Treasury Dept. Overseas trade report. 1951-	
	1953, 1962-63	382.72972 D671 f

CA	DOMINICA. Import and export statistics. 1920-	

CA	DOMINICA. Treasury Dept. Export trade. Issued by the statistics section of the Dept.

CB	DOMINICA. Treasury Dept. Monthly overseas trade report. Issued by the statistics section of the Dept.

AA	WEST INDIES. Federal Statistical Office. Agricultural statistics, series 1. (1956-58 Survey)
	FRI no. 4, The survey in Dominica, 1960

AR	DOMINICA. Dept. of Agriculture and Forestry. Report. 1911/12-	
	1960+	S183 D6A3

DOMINICAN REPUBLIC

SA	DOMINICAN REPUBLIC. Dirección General de Estadística. Anuario estadístico de la República Dominicana. 1936-54//	
	1938, 1942-54	HA886 A35

SA CA	GREAT BRITAIN. Dept. of Overseas Trade. ... Report on the economic, financial and commercial conditions in the Dominican Republic ... and in the Republic of Hayti .... [1921-38.] Title varies.	
	1921, 1923-25, 1928, 1930, 1932-33, 1936, 1938	380.005 G786

SA	**DOMINICAN REPUBLIC. Dirección General de Estadística. Informe. 1906, 1911- (?)

SB	DOMINICAN REPUBLIC. Dirección General de Estadística. Boletín de divulgaciónes. no. 1, 1958-	
	no. 1, 1959-no. 6, 1960	330.97293 D6714

SB	DOMINICAN REPUBLIC. Dirección General de Estadística. Informaciónes estadísticas dominicanas. v. 1, no. 1-v. 3, no. 1; Nov. 1951-Mar. 1954.	
	v. 1, no. 1-5; v. 3, no. 1; Nov. 1951-Mar. 1954	HA886 A4

CA	**DOMINICAN REPUBLIC. Oficina de Contralor y Receptor General de las Aduanas. Sumario del comercio ... Summary of commerce. 1905- (?) Spanish and English.

CB CA	DOMINICAN REPUBLIC. Dirección General de Estadística y Censos. Comercio exterior. 1953- 　Published irregularly, usually 12 nos. in 3-4 issues. Annual cumulation in final issue each year since 1954. 　Vol. 6, no. 4-11, not published.

v. 1, 1953-v. 11, 1963+   382.7293
wanting: Aug. 1953,   D671c
  Oct., Nov. 1954

----- ----- Suplemento de comercio exterior de la República Dominicana. Datos relativos al período 1950-1962.

1964   382.7293
                                    D761a

CB DOMINICAN REPUBLIC. Dirección
CA   General de Estadística.
       Exportación de la República Dominicana. (?) -1952//
      Superseded by its Comercio exterior.

1941, 1944, 1945,   382.7293
  1947-52   D671e

CB DOMINICAN REPUBLIC. Dirección
CA   General de Estadística.
       Importación de la República Dominicana. (?) -1952//
      Superseded by its Comercio exterior.

1943-52   382.7293
                                    D671i

CB **DOMINICAN REPUBLIC. Oficina de Contralor y Receptor General de las Aduanas.
      Sumario trimestral del comercio ... Quarterly summary of commerce. 1-2, Apr.-July 1905//

AR DOMINICAN REPUBLIC. Secretaria de Agricultura.
      Memoria.

1909, 1925, 1926, 1928   630.67293
                                         D672

DUTCH EAST INDIES

see also INDONESIA

SA DUTCH EAST INDIES. Centraal Kantoor voor de Statistiek.
      Statistical pocket book of Indonesia.
      Title varies: 1940, Pocket edition of the statistical abstract of the Netherlands Indies.

1940-41   HA1811
                                        A322

SA DUTCH EAST INDIES. Centraal Kantoor voor de Statistiek.
      Statistisch jaaroverzicht van Nederlandsch-Indië. [1-2]- Jaarg. 1922-23--
      Beginning with 1930 (8. jaarg.) the publication forms [deel] II of Indisch verslag, issued by Netherlands (Kingdom, 1815-), Departement van Koloniën (325.2492 N469i).

1922/23-1940   319.1
wanting: 1939   D975s

SA DUTCH EAST INDIES. Departement van Landbouw, Nijverheid en Handel.
      Handbook of the Netherlands East Indies. 1916-
      1916-20 have title: Yearbook of Netherlands East Indies.

1916, 1920, 1924, 1930   319.1
                                         D975

SB DUTCH EAST INDIES. Centraal Kantoor voor de Statistiek.
      ... Mededeelingen van het Centraal Kantoor voor de Statistiek. Bulletin of the Central Bureau of Statistics. 1920-

no. 19, 30, 36, 39, 44,   319.1
  46, 51, 60, 69, 74, 96;   D973
  1925-31

CB DUTCH EAST INDIES. Centraal Kantoor voor de Statistiek.
      Maandstatistiek van den in- en uitvoer van Nederlandsch-Indië.

1940: 1, 4, 9-10, 12;   HF247
  1948: 5   A48

AA	DUTCH EAST INDIES. Departement van Landbouw, Nijerheid en Handel. *Jaarboek.* 1906-				
	1909-11, 1913-14, 1919-21	630.691 D975			

DUTCH GUIANA

see SURINAM

E

EAST AFRICA

see also KENYA, TANGANYIKA, TANZANIA, UGANDA

SA East African Common Service Organization.
Report.
1948-56 part of numbered series of Gt. Brit. Colonial Office. no. 245: 1948-no. 331: 1956 (325.342 G784).

1957-64+	325.342 G7871AfE
FRI 1957-64+	

SA East African Common Services Organization. East African Office, London.
Statistical digest.

1961-62	HA2001 E2

SA East African Common Services Organization. East African Statistical Dept.
Report.
Report for 1950-1960/61 published under the Authority of the East Africa High Commission.

1950-1961/62 wanting: 1960/61	316.76 E13

SA CA GREAT BRITAIN. Dept. of Overseas Trade.
... Economic conditions in East Africa and in Northern Rhodesia and Nyasaland ... Report .... 1921-38.
Includes Uganda Protectorate, Kenya Colony and Protectorate, Tanganyika Territory, and Zanzibar Protectorate.
Title varies.
1929/30 includes report for British Somaliland. 1932-34 and 1936/37 include annex on the Somaliland Protectorate.

1921-38	380.005 G786
HL 1921-28, 1934/36-1937/38	HC517 E2G78

SB East African Common Services Organization. East African Statistical Dept.
Economic and statistical review. Dec. 1961-
Supersedes the *Economic and statistical bulletin* of the East African Statistical Dept. of the East Africa High Commission.

Dec. 1961-June 1965+	330.967606 E108ec
FRI 1961-65+	

SB East Africa High Commission. East African Statistical Dept.
Economic and statistical bulletin.
no. 1-52, Sept. 1948-June 1961//
Superseded by the *Economic and statistical review* of the East African Statistical Dept. of the East African Common Services Organization.

no. 10-52 wanting: 11-14	330.967606 E108e
FRI no. 22-52	

CA East African Common Services Organization. East African Customs and Excise Dept.
Annual trade report of Kenya, Uganda and Tanganyika.
Issued through 1948 by the Commissioner of Customs, Kenya and Uganda; 1950-60 by the East Africa High Commission.

1925, 1941-64+	382.676 K37 f

CA GREAT BRITAIN.
    Report by His Majesty's Commissioner on the East Africa Protectorate.

    1901   Cd. 769, Brit. Doc. 1901
            xlviii
    1903   Cd. 1626, Brit. Doc. 1903
            xlv
    1903-04 Cd. 2684-21, Brit. Doc. 1906
            lxxiii
    1905-06 Cd. 3285-6, Brit. Doc. 1907
            liii
    1906-07 Cd. 3729-21, Brit. Doc. 1908
            lxviii
    1907-08 Cd. 4448-1, Brit. Doc. 1909
            lvii

CB East African Common Services Organization. East Africa Customs and Excise Dept.
    Trade and revenue report for Kenya, Uganda and Tanganyika. Monthly.

AR EAST AFRICA PROTECTORATE. Dept. of Agriculture.
    Annual report.
    Continued by the Annual report of the Dept. of Agriculture of Kenya Colony and Protectorate.

    1911/12-1917/18   630.6676
                                 E13

ECUADOR

SA ECUADOR. Dirección General de Estadística y Censos.
    Síntesis estadística del Ecuador. (?) -1955/62//
    Vol. for 1955/58 issued as Anexo al informe del Ministerio de Economía.

    1955/58   HA1022
                                 A55

SA ECUADOR. Dirección General de Estadística.
    Ecuador en cifras, 1938 a 1942.
    Compilation of annual data.

    1938/42   318.41
                                 E19e
    FRI 1938/42

SA GREAT BRITAIN. Dept. of Overseas
CA   Trade.
    ... Economic and financial conditions in Ecuador .... [1921-38.]
    Title varies.

    1921-23, 1925, 1928,   380.005
        1930, 1934, 1936, 1938   G786

SB ECUADOR. Dirección General de Estadística y Censos.
    Boletín estadístico. 1, Dec. 1954/Jan. 1955- (?)
    No longer published.

SB ECUADOR. Dirección General de Estadística y Censos.
    El trimestre estadístico del Ecuador. May 1945-Apr. 1947.

SB ECUADOR. Dirección General de Estadística y Censos.
    Cuadernos de estadística. 1945-
    Quarterly.

    1945: no. 1-2   318.31
                                 E19

CA ECUADOR. Dirección de Financiamiento y Asesoría Fiscal.
    Comercio exterior ecuatoriano. 1957-
    Cover title: Anuario de comercio exterior.
    Vols. for 1957-62 issued by various agencies in the Ministerio del Tesoro.

    1962-64+   382.841
                                 E192
                                 f

CA GREAT BRITAIN. Board of Trade.
    Ecuador; economic and commercial conditions in Ecuador.
    Issued by the Commercial Relations and Exports Dept. in its Overseas economic surveys series.

    1950, 1954   380.005
                                 G786aec

CA   ECUADOR. Dirección General de Estadística.
     ... Comercio exterior de la República del Ecuador.
     Title and issuing agency vary.

     1916/25-1925/26   382.841
                                   E19

CA **ECUADOR. Dirección de Estadística Comercial.
     Resumen estadístico del comercio exterior. 1911/20//

CA **ECUADOR. Dirección de Estadística Comercial.
     Estadística comercial de la importación y exportación por las aduanas de la República. 1889-1905(?)//
     Early years as Anuario de estadística comercial.

CB **ECUADOR. Ministerio de Hacienda.
     Boletín estadístico del movimiento comercial. no. 1-9, 1908/09-1915/16//
     No. 1 as Estadistica fiscal; 2, Boletín de estadística fiscal y comercial; 3-8, Boletín estadístico comercial.

AB **ECUADOR. Dirección General de Agricultura.
     Boletín. no. 1-4, 1921//(?)

EGYPT

see also UNITED ARAB REPUBLIC

SA   EGYPT. Ministry of Finance and Economy. Statistical Dept.
     Annuaire statistique de Poche. 1940-
     Published also in English.

     1946-51   316.2
                                   E32p

SA **EGYPT. Statistical Dept.
     Tableaux statistiques ... 1881-97. 1, 1898//

SA **EGYPT. Statistical Dept.
     Statistique de l'Egypte ... 1863-72. 1, 1873//

CA **EGYPT. Customs Administration.
     Le commerce extérieur de l'Egypte. Statistique comparée 1884-1903. 1, 1904//

CA **EGYPT. Customs Administration.
     Le commerce extérieur de l'Egypte. Statistique comparée 1884-89, accompagnée de tables graphiques et tableaux comparatifs depuis l'année 1874. 1, 1891//

ERITREA

see also ETHIOPIA

SB   ERITREA.
CA   No statistical bulletin of Eritrea available.
     Some statistical data appeared in the following: Italy. Bolletino statistico dell'Africa italiana. Jan. 1938-Oct. 1939. Monthly, no longer published.
     See Ethiopia (CA) for some statistical data before federation with Ethiopia.

CA   ITALY. Ministero dell'Africa Italiana. Ufficio Studi e Propaganda.
     ... Statistica del movimento commerciale marittimo dell'Eritrea, della Somalia italiana, della Tripolitania e della Cirenaica e del movimento commerciale carovaniero dell'Eritrea .... 1921 e 1922-
     1925-26 includes "Movimento commerciale marittimo dell'Oltre Giuba dal 1° luglio 1925 al 20 giugno 1926."
     1928-29 includes "Movimento della navigazione marittima della quattro colonie."

     1923-24   382.45
                                   I892

     HL 1923-24, 1931-32   HF270
                                   I88

ESTONIA

SA   GREAT BRITAIN. Dept. of Overseas
CA   Trade.
        ... Report on economic and commercial conditions in Estonia .... [1923-37.]
        Title varies.

    1923, 1925, 1926, 1929,   380.005
    1932, 1935, 1937   G786

SA   ESTONIA. Riigi Statistika Keskbüroo.
        Estonie de 1920-1930. Résumé rétrospectif.

    HL   1920-30 (in 1 vol.)   HC337
                                       E7A54

SB   ESTONIA. Riigi Statistika Keskbüroo.
        Eesti statistika kuukiri, recueil mensuel du Bureau Central Statistique de l'Estonie. 1922-

    1922: 1-11; 1923: 12,   314.74
       13, 17   E79

CA   ESTONIA. Riigi Statistika Keskbüroo.
        Eesti majandus. Statistique économique. v. 1, 1923-

    HL   v. 1-14, 1923-31

CA   ESTONIA. Riigi Statistika Keskbüroo.
        Väliskaubandus. Commerce extérieur de l'Estonie. 1921-

    HL   1921-22

AA   ESTONIA. Riigi Statistika Keskbüroo.
        Eesti pollumajandus. Annuaire de la statistique agricole. 1922-

    HL   v. 2, 1924; v. 4-6,
           1925-27

ETHIOPIA

SA   ETHIOPIA. Central Statistical Office.
        Statistical abstract. 1963-

    1963-64+   HA1961
                                   A3
                                   f

SA   ETHIOPIA. Ministry of Commerce and Industry.
        Economic progress of Ethiopia. 1924-54, 1955.

    1955   330.963
                                   E84e

SA   ETHIOPIA. Ministry of Commerce and Industry.
        Economic handbook of Ethiopia. 1951, 1958.

    1951   330.963
                                   E84

SA   GREAT BRITAIN. Dept. of Overseas
CA   Trade.
        ... Report on economic conditions in Ethiopia ..., 1929-1931.

    1929-31   380.005
                                   G786

SB   ETHIOPIA. Imperial Planning Board.
        Monthly statistical bulletin. Mar. 1959-   Irregular.
        Ceased publication with issue for Nov. 1963.

CA   ETHIOPIA. Central Statistical Office.
        Summary report on Ethiopia's external trade. 1953-63.
        "Ethiopia's imports and exports in commodity detail, 1962 (covering the period Jan. 11, 1962 through Jan. 10, 1963": 16 l. at end.

    1953-63   HF3936
                                   A4
                                   f

CA   ETHIOPIA. Statistical Dept.
       Report on the external trade.
       Report for 1949/53 includes statistics for Eritrea before federation with Ethiopia.

   1946-52, 1949-53,   382.63
     1949-55, 1954-55,   E88
     1955-56   f

CA   ETHIOPIA. Customs Adviser's Office.
       Import and export trade statistics.
     1955/56-
       Continues its Import trade statistics and Export trade statistics.

   1955/56   382.63
                                  E86ie

CA   ETHIOPIA. Customs Adviser's Office.
       Export trade statistics.
       Continued in its Import and export trade statistics, 1955/56-

   1954/55   382.63
                                  E86e
                                  f

CA   ETHIOPIA. Customs Adviser's Office.
       Import trade statistics.
       Continued in its Import and export trade statistics, 1955/56-

   1954/55   382.63
                                  E86i
                                  f

CB   ETHIOPIA. Ministry of Commerce and Industry.
       Ethiopian economic review. no. 1, Dec. 1959-   Irregular, quarterly.

   no. 1, Dec. 1959-no. 5,   330.963
     1962   E842

CB   ETHIOPIA. Ministry of Commerce and Industry.
       Bulletin. 1952-Apr. 1959.
       Title varies; no. 1-10, 1952-57, Quarterly Bulletin.

   v. 3, no. 1-v. 6, no. 1:   HC591
     Aug. 1954-Feb. 1957;   A3A5
     no. 11-no. 15: July
     1957-59

CB   ETHIOPIA. Customs Adviser's Office.
       Quarterly import and export trade statistics.

AB   ETHIOPIA. Ministry of Agriculture. Division of Agricultural Economics and Statistics.
       Quarterly exports of principal agricultural products. Statistical bulletin. 1, May 1955-

F

FALKLAND ISLANDS

SA   GREAT BRITAIN. Colonial Office.
       Annual report on the Falkland Islands and dependencies. 1947-
       Previously issued in its numbered series, Colonial reports--annual (325.342   G787) no. 1076: 1919- no. 1932: 1938. Suspended 1940-46.

   1947-63+   325.342
                                  G7871Fa

SA **FALKLAND ISLANDS.
       Blue book. 1846-(?)

Federation of RHODESIA AND NYASALAND

see RHODESIA AND NYASALAND

see also

RHODESIA (NORTHERN AND SOUTHERN)
RHODESIA, NORTHERN   also ZAMBIA
RHODESIA, SOUTHERN   also RHODESIA
NYASALAND   also MALAWI

Federation of SOUTH ARABIA

see also ADEN

CA FEDERATION OF SOUTH ARABIA.
     Ministry of Commerce and Industry.
     Statement of external trade. 1955-56--
     Issued 1955-56--1959 by Trade Dept.,
     Aden; 1960-61, by Customs and Excise
     Dept., Aden; 1962, by Ministry of Com-
     merce and Industrial Development, Fed-
     eration of South Arabia. Supersedes
     Aden's Trade and navigation report.

   1956/57-1962+   HF240
                            S6A3
                            f

CB FEDERATION OF SOUTH ARABIA.
     Ministry of Commerce and Industry.
     Trade bulletin.

   Jan.-June 1964+   HF240
                            S6A4

AR FEDERATION OF SOUTH ARABIA.
     Dept. of Agriculture.
     Report.
     Continues the report formerly issued
     by the Dept. of Agriculture of the Aden
     Protectorate.

   1958-63 (1 vol.)   S322
                            S6A3

FIJI

SA GREAT BRITAIN. Colonial Office.
     Annual report on Fiji. 1946-
     Previously issued in its numbered
     series, Colonial reports--annual
     (325.342 G787) no. 1080: 1920-
     no. 1910: 1938. Suspended 1940-46.

   1947-64+   325.342
                            G7871Fi

SA FIJI.
     Annual report on Fiji.

   1949-50   354.9611
                            F477r

SA FIJI.
     Fiji blue book. 1914-40(?)

   1914, 1916-18, 1923-38, 354.9611
     1940   F477

SB FIJI. Commerce and Industries Office.
     Quarterly statistical summary.
     Compiled by the Government Statisti-
     cian.

   Mar./June 1963-   HA4007
     Oct./Dec. 1965+   F5A4
                            f

CA FIJI. Comptroller of Customs.
     Trade report.

   1951-64+   382.9611
                            F477
                            f

CA **FIJI. Customs Dept.
     Report on the trade, commerce and
     shipping.
     Early reports as Statement of the
     trade and navigation of the Colony. 1899-
     1927.

CB FIJI. Commerce and Industries Office.
     Quarterly digest of trade statistics.

   Sept. 1965+   HF293
                            F5A4

AR FIJI. Dept. of Agriculture.
     Annual report. 1910-

   FRI 1916-63+
        wanting: several issues

FINLAND

SA FINLAND. Statistiska Centralbyrån.
     Statistisk årsbok för Finland. n.s. 1,
     1903-

   1909-64+   314.71
                            F511s

SA   FINLAND. Statistiska Centralbyrån.
     Ofversigt af Finlands ekonomiska
tillstand. 1876-80. 1 vol.
     Issued as Bidrag till Finlands officiela
statistisk.

1876-80   330.9471
                           F511

SB   FINLAND. Statistiska Centralbyrån.
     Tilastokatsauksia ... Statistiska
översikter 1924-
     Finnish, Swedish, and French, 1949-
51; Finnish, Swedish, and English, 1952-

v. 24: 9/10, Sept./Oct.   314.71
1949-Sept. 1965+   F511t

CA   FINLAND. Tullihallituksen Tilastotoi-
     mistossa.
     Ulkomaankauppa vuosijulkaisu.
Utrikeshandel årspublikation. Commerce
extérieur. 1903-
     Continues in part its Handel och
sjöfart, assuming its voluming.
     Title varies.

1909-26, 1932-33,   382.471
  1935-36, 1940-59+   F511a

FRI Current two years
     only.

CB   FINLAND. Tullihallituksen Tilastotoi-
     mistossa.
     Ulkomaankauppa kuukausijulkaisau.
Utrikeshandel månadspublikation. Commerce extérieur de la Finlande.
     Title varies.

1909-27, 1944-Jan. 1965+   382.471
wanting: Oct.-Dec. 1910,   F511
  Mar.-Apr. 1911, Mar.
  1912, July, Dec. 1913,
  Sept., Oct., Dec. 1926

AA   FINLAND. Taliarollisessa Päätoimis-
     tossa.
     Maatalous: maatalouden vuositilasto.
Agriculture: annual statistics.

FRI 1951-64+

AA   FINLAND. S. Virallinen Tilasto
     Maatalous.
     Maanviljelys ja karjanhorto vnonna.
L'Agriculture et l'élevage du betail.

FRI 1939-49, 1959+

AA   FINLAND. Lantbruks Styrelsen.
     Suomen maataloutta kuvioin ja kartoin.
     L'Agriculture de la Finlande en dia-
grammes et cartes.

HL   1925   HD1995.5
                           A35
                           f

FORMOSA

see also CHINA (TAIWAN)

SA **FORMOSA.
     Annual statistics .... 1, 1900-(?)
In Japanese.

CA   FORMOSA. Dept. of Statistics.
     Taiwan trade statistics for the last
fifty-three years (1896-1948).

1896-1948 (1 vol.)   382.5142
                           F726

CA   FORMOSA.
     ... Returns of the trade of Taiwan
(Formosa) for the twenty-five years from
1896 to 1920 inclusive.

1896-1920 (1 vol.)   382.512
                           F726

CA **FORMOSA.
     Annual return of the trade of Taiwan
(Formosa). 1897-   In Japanese and
English.

FRANCE

SA   FRANCE. Institut National de la Statistique et des Etudes Economiques.
<u>Annuaire statistique</u>. 1878-
Suspended 1940-45, 1947-51.

1878-1963+	HA1212
wanting: 1955, 1960	A4

SA   FRANCE. Institut National de la Statistique et des Etudes Economiques.
<u>Annuaire statistique de la France. Rétrospectif</u>. Irregular.

FRI n. s. no. 8, 1961

SA   FRANCE. Statistique Générale.
<u>Annuaire statistique abrégé</u>. 1943, 1949.

1943, 1949	314.4
	F814ab

SA   FRANCE. Statistique Générale.
... <u>Statistique générale de la France. Historique et travaux de la fin du XVIII siècle au début du XX avec 103 tableaux graphiques relatifs aux travaux les plus récents</u>.
"Publications de la Statistique générale de la France de l'origine à 1913."

1913	314.4
	F814sg

SB   FRANCE. Institut National de la Statistique et des Etudes Economiques.
<u>Bulletin mensuel de statistique</u>. Oct. 1911-
Suspended Oct. 1939-May 1941.

v. 14-37, n.s. année	314.4
1-11, 13-15; Oct.	F814b
1924-Dec. 1949,	
Jan. 1950-Dec. 1964+	
wanting: several issues	

SB   FRANCE. Institut National de la Statistique et des Etudes Economiques.
<u>Etudes statistiques, supplément</u> [to <u>Bulletin mensuel de statistique</u>]. 1932-
Discontinued Dec. 1964. Continued in part in its <u>Etudes et conjoncture</u>.

1932-39 (monthly), 1948-	314.4
49, n.s. 1950-54,	F814b
1958-60, 1962-64	Supp.

SB   FRANCE. Institut National de la Statistique et des Etudes Economiques.
<u>Bulletin hebdomadaire de statistique</u>.

HL   no. 36-860: Jan. 8, 1949-Dec. 26, 1964
wanting: no. 174: Sept. 1951; no. 312: Apr. 1954

CA   FRANCE. Direction Générale des Douanes et Droits Indirects.
<u>Statistiques du commerce extérieur de la France; annuaire abrégé</u>. 1963-
Title varies: 1963, <u>Annuaire abrégé de statistiques: commerce extérieur</u>.

1963-64+	HF193
	A238

CA   FRANCE. Direction Générale des Douanes.
<u>Tableau général du commerce extérieur</u>.
Title varies: 1825-95, <u>Tableau général du commerce de la France avec ses colonies et les puissances étrangères</u>; 1896-1927, <u>Tableau général du commerce et de navigation</u>.

1844, 1883, 1914: v. 1;	382.44
1917: v. 1-2; 1933-36	F81

CB   FRANCE. Direction Générale des
CA   Douanes et Droits Indirects.
<u>Statistiques du commerce extérieur de la France</u>.
Annual 1939-59; 1960- quarterly with cumulative Dec. issue.

1960: 1-2; 1961: 1, 4;	HF193
1964-65+	A23

CB FRANCE. Direction Générale des
   Douanes et Droits Indirects.
  Statistiques du commerce extérieur de
la France en nomenclature C. T. C. I.

  1964+   HF193
           A232

CB FRANCE. Direction Générale des
   Douanes et Droits Indirects.
  Statistiques du commerce extérieur.
Commentaires mensuels. Jan. 1961-
  Supersedes in part its Statistique mensuelle du commerce extérieur de la France and its Tableau général du commerce extérieur: commerce de la France avec ses colonies et les pays étrangers.

  1961-   HF193
           A24

CB FRANCE. Direction Générale des
   Douanes et Droits Indirects.
  Statistique mensuelle du commerce extérieur de la France.
  Issues for Feb. 1938-1941 published in two parts.
  Title varies: 1863-1923, Documents statistiques sur le commerce de la France.

  1912-59   382.44
  wanting: several issues; F81d
  many years have Dec.
  issue only

AA FRANCE. Direction de l'Agriculture.
  . . . Statistique agricole annuelle.
  Title varies.

  FRI 1903, 1906-20,
     1922-24, 1926,
     1928-35, 1943-63+

AA FRANCE. Direction des Etudes et du
   Plan.
  Statistique agricole (rétrospectif 1930-1957).
  Occasionnel.

  1930-57 (1 vol.) 630.944
           F815st

  FRI 1930-57 (1 vol.)

AB FRANCE. Direction Générale des Etudes
   et des Affaires Générales.
  Statistique agricole. no. 1, Apr.
1962-

  FRI no. 1, 1962-Apr.
     1966+

AB FRANCE. Direction Générale des Etudes
   et des Affaires Générales.
  Cahiers mensuels de statistique agricole. no. 65, June 1962-
  Supersedes Bulletin mensuel de statistiques agricoles.
  Supplement: Résultats provisoires de la statistique agricole. Published in Jan. of each year with provisional figures for preceding year.
  Supplement: Résultats sommaires de la statistique agricole. Published in July of each year with summary figures for preceding year.

  FRI 1962-66+
    Supplement:
     Provisoires, 1963-64
     Sommaires, 1961-62, 1964

AB FRANCE. Direction Générale de l'Agriculture.
  Bulletin mensuel de statistiques agricoles. Nov. 1956-May 1962.
  Superseded by Direction Générale des Etudes et des Affaires Générales. Cahiers mensuels de statistique agricole.

  FRI no. 57-64, 1961-62

AB FRANCE. Ministère de l'Agriculture.
  Revue du Ministère de l'Agriculture.
1945-59//
  Superseded in 1963 by Revue française de l'agriculture. Does not include statistics.

  FRI July 1946-59

FRANCE (Overseas)

SA   FRANCE. Institut National de la Statis-
         tique et des Etudes Economiques.
         Annuaire statistique des territoires
d'outre-mer. 1939-
        Vols. for 1939/49--1949/54 issued by
the Ministère de la France d'Outre-Mer,
Service des Statistiques, in collaboration
with the Institut.
        Title varies.
        Some vols. issued in 12 parts, each
on specific subjects.

1939/46, pt. F; 1938/48, pt. J; 1939/49, v. 1 and 2; 1949/54, fasc. I; 1949/55, fasc. II; 1959-61--	325.344 F817
HL 1939/46, pts. A-B, E-F, I-K	HA1221 A4
1949/54, fasc. I	A42
1949/55, fasc. II	HA1228 A24

SA   FRANCE. Service des Statistiques
         d'Outre-Mer.
         Outre-mer 1958; tableau économique
et social des Etats et territoires d'outre-
mer a la veille de la mise en place de
nouvelles institutions.

1958	325.344 F8177
FRI 1958	

SA   FRANCE. Ministère de la France
         d'Outre-Mer. Service des Statis-
         tiques.
         Inventaire social et économique des
territoires d'outre-mer. 1950-55.

1950-55	325.344 F8173
FRI 1950-55	

SA   FRANCE. Ministère de la France
         d'Outre-Mer. Service des Statis-
         tiques.
         Annuaire statistique des possessions
françaises: années antérieures à la
guerre. Ed. provisoire, 1944-46.
        1 vol. (looseleaf).

1944-46	HA1228 A22
HL 1944-46	HA1219 A4

SA   The general statistics of the overseas
territories of France were included with
the Annuaire statistique (314.4   F814a)
for the years 1878-1938.

SA   **FRANCE. Ministère des Colonies.
         Statistiques coloniales. 1837-96//
        1839 as Tableaux et relevés de popula-
tion, de culture, de commerce, de navigation, [etc.] ... la suite des tableaux
et relevés insérés dans les Notices statistiques sur les colonies françaises;
1840-81, Tableaux de population, de culture, de commerce, et de navigation,
[etc.]. 1860-70 also published in or as
supplements to Revue maritime. 1892-95
in 1 vol. as Résumé des statistiques
coloniales.
        Continued by the following publications
issued by Agence Générale des Colonies:
Statistiques de l'industrie minière dans
les colonies françaises; Statistiques de
la population; Statistiques de la navigation; Statistiques des chemins de fer;
Statistiques des finances; Statistiques du
commerce.

SB   FRANCE. Institut National de la Statis-
         tique et des Etudes Economiques.
         Données statistiques. 1961-

n.s. no. 1, 1961-Dec. 1965+	314.4 F814d

SB   FRANCE. Institut National de la Statis-
         tique et des Etudes Economiques.
         Résumé des statistiques d'outre-mer.
Bulletin accéléré. no. 1, Oct. 1958-

     FRI   no. 1, Oct. 1958-
           no. 64, Dec. 1964+

SB FRANCE. Institut National de la Statistique et des Etudes Economiques.
    Bulletin mensuel de statistique d'outre-mer. v. 1, Sept. 1945-
    Supersedes Bulletin de statistiques coloniales issued by the Service Colonial des Statistiques.
    Title varies.
    Monographic supplements issued irregularly in two series:
        Supplément. Série statistique.
        Supplément. Série études.

    v. 1-16: no. 4, Sept. 1945-   HA1228
        Dec. 1960   A24
    Série études: no. 1, 12-16,
        18-25, 27-28, 30-36
    Série statistique: no. 1-3,
        5-13, 15-20

CA FRANCE. Institut National de la Statistique et des Etudes Economiques.
    Compendium des statistiques du commerce extérieur des pays africains et malgache. 1938-

    1949-63+   382.44
    wanting: 1960   F815c

CA FRANCE. Institut National de la Statistique et des Etudes Economiques.
    Statistiques du commerce extérieur des territoires d'outre-mer.
    Title varies.

    1938, 1945, 1957, 1959,   382.44
        1961+   F815

    HL 1945

CA FRANCE. Institut National de la Statistique et des Etudes Economiques.
    Commerce extérieur des états d'Afrique et de Madagascar de 1949 à 1960. Rétrospectif.

    FRI 1949-60

CA FRANCE. Office Colonial. Ministère des Colonies.
    Statistiques du commerce des colonies françaises. 1897-1914.
    Continues in part Statistiques Coloniales. 1897-1905 as Statistiques Coloniales. Commerce. Continued as Renseignements généraux sur le commerce.

    1906, v. 1-2; 1913,   382.44
        v. 1-2; 1914, v. 1   F814

CA **FRANCE. Agence Générale des Colonies.
    Renseignements généraux sur le commerce des colonies françaises. 1914/17//
    Continues Statistiques du commerce and Statistiques de la navigation. Continued as Renseignements généraux sur le commerce et la navigation ....

CA **FRANCE. Agence Générale des Colonies.
    Renseignements généraux sur le commerce et la navigation des colonies françaises. 1918-(?)

CB FRANCE. Ministère de la France d'Outre-Mer.
    Résumé du commerce extérieur des territoires d'outre-mer.

    1956   HF193
                                    A6

FRENCH EQUATORIAL AFRICA

after August 1960 see

  CENTRAL AFRICAN REPUBLIC
  CHAD
  CONGO (BRAZZAVILLE)
  GABON

SA FRENCH EQUATORIAL AFRICA. Service de la Statistique Générale.
    Annuaire statistique de l'Afrique équatoriale française. 1936/50-

    1936/50   316.72
                                    F877

    FRI 1936/50, 1951-55

SA **FRENCH EQUATORIAL AFRICA.
   Annuaire de l'Afrique équatoriale française. 1906-22//(?)
   1906-11 as Annuaire officiel de la Colonie du Congo ....
   Title varies slightly.

SB FRENCH EQUATORIAL AFRICA. Service de la Statistique Générale.
   Bulletin mensuel de statistique.
   Issued 1959 by the agency under a variant name: Bureau Central de la Statistique.
   Superseded by Bulletin des statistiques générales, issued by Union douanière équatoriale.

   Jan. 1957-June 1959,   HF268
   Sept./Oct. 1959   E7A33

CA FRENCH EQUATORIAL AFRICA. Service de la Statistique Générale.
   Statistiques du commerce extérieur.
   Vols. for 1950-52 issued in two parts: 1, Exportations; 2, Importations.
   Superseded by Commerce extérieur, issued by Union douanière équatoriale.

   1950-59   HF268
                                   E7A32

CA FRANCE. Ministère de la France d'Outre-Mer. Service des Statistiques.
   Statistiques sommaires du commerce extérieur de l'Afrique équatoriale française.
   Superseded by Service de la Statistique Générale. Statistique du commerce extérieur.

   1936-46, 1947-49   382.44
                                   F815s

   FRI 1936-46, 1947-49

CA FRENCH EQUATORIAL AFRICA.
   Statistiques douanières pendant les années ....

   HL 1936/38-1942/44

FRENCH GUIANA

SA FRANCE. Institut National de la Statistique et des Etudes Economiques.
   Annuaire statistique de la Guyane. 1947/52-

   1947/52-1957/59   HA1037
                                   F7A3

SA **FRENCH GUIANA.
   Annuaire de la Guyane française. 1861-(?)
   Continues Almanach de la Guyane française.
   1920/22 issued in 1930.

SA **FRENCH GUIANA.
   Almanach de la Guyane française. 1820-60//
   Continued as Annuaire de la Guyane française.

FRENCH GUINEA

see GUINEA, FRENCH
   1958- GUINEE (REPUBLIQUE)
see also FRENCH WEST AFRICA

FRENCH SUDAN

see SUDAN, FRENCH
   1960- MALI
see also FRENCH WEST AFRICA

FRENCH WEST AFRICA

Consisted of DAHOMEY, DAKAR, FRENCH GUINEA, FRENCH SUDAN, HAUTE VOLTA, IVORY COAST, MAURITANIA, NIGER, and SENEGAL. See also these entries (except Dakar).

SA FRENCH WEST AFRICA. Service de la
    Statistique Générale.
    Annuaire statistique de l'Afrique occi-
    dentale française. 1933/34-
    Continues yearbook by the same title
    published by the Agence Economique
    1933-38.

    1936-38, 1949, 1951: 2,   HA2101
      1950-54: 1-3,   F7
      1955-56-57: 1

SA FRANCE. Haut Commissariat en Afrique
    Occidentale Française.
    A.O.F. 1957; tableaux économiques.

    1957   HC547
                                 W5A53

    FRI 1957

SA GREAT BRITAIN. Dept. of Overseas
CA   Trade.
    ... Economic conditions in French
    West Africa .... 1928-36.

    1928-36   380.005
                                 G786

SA **FRENCH WEST AFRICA.
    Annuaire du gouvernement général de
    l'Afrique occidentale française. 1903-
    22//
    Continues Senegal. Annuaire du Séné-
    gal et dépendances.

    1917/21, 1922   316.61
                                 F877

SB FRENCH WEST AFRICA. Service de la
    Statistique du Sénégal et de la Mauri-
    tanie.
    Bulletin statistique.
    Superseded in part by: Sénégal. Ser-
    vice de la Statistique et de la Mécanogra-
    phie. Bulletin statistique et économique
    mensuel.

    1955: 1-5, 1956: 1-6, 1957   HA2117
    wanting: Apr. 1957   S4F7

SB FRENCH WEST AFRICA. Direction
    Générale des Services Economiques.
    Bulletin économiques mensuel.

SB FRENCH WEST AFRICA. Service de la
    Statistique Générale.
    Bulletin de la statistique générale de
    l'Afrique occidentale française.

CB FRENCH WEST AFRICA. Direction
CA   Fédérale des Douanes.
    Statistiques du commerce et de la
    navigation.

    1952: Année I and II   HF267
    1953: Q. 1st, 2d (1 and 2), A3
      3d (1 and 2); Année II
    1954: Q. 1st, 2d, 3d
      (1 and 2); Année I and II

CB FRENCH WEST AFRICA. Service de la
    Statistique Générale.
    Statistiques du commerce extérieur de
    l'Afrique occidentale française.
    Supersedes Gouvernement Général de
    l'Afrique Occidentale Française. Statis-
    tiques mensuelles du commerce extérieur
    de l'A.O.F.--Commerce spécial--Impor-
    tations--Exportations.

    1940-51   HF267
    wanting: several issues   A35
                                 f

AR FRENCH WEST AFRICA. Direction
    Générale des Services Economiques.
    Inspection Générale de l'Agriculture.
    Rapport annuel.

G

GABON

see also FRENCH EQUATORIAL AFRICA
    before 1960

SB Union Douanière Equatoriale.
    Bulletin des statistiques générales.
    no. 1, Jan. 1963-
    Supersedes Bulletin mensuel de statis-
    tique formerly issued by Service de la
    Statistique Générale, French Equatorial
    Africa.

    no. 1, Jan. 1963-   HF268
      no. 13, Jan. 1966+   E7A52

SB   GABON. Service des Statistiques.
      Bulletin mensuel de statistique. v. 1,
Apr. 1959-

      v. 4, no. 34-v. 7, no. 71;
Jan. 1962-Feb. 1965+

CA   Union Douanière Equatoriale.
      Commerce extérieur, U.D.E. et
républiques composantes: Centrafrique,
Congo, Gabon, Tchad.
      Supersedes Statistiques du commerce
extérieur formerly issued by Service de
la Statistique Générale, French Equatorial Africa.

      1960-63+   HF268
                               E7A5

CA   GABON. Service des Statistiques.
      Statistiques du commerce extérieur.
Issued as supplement to its Bulletin
mensuel de statistique.

GAMBIA

SA   GAMBIA.
      Statistical summary. Annual. 1964-
Issued in the series of Sessional
papers.
      Summary for 1962-63 was issued as
part of the Report of the Gambia (London),
1962-63.

      1964+   HA2275
                               G3A3

SA   GREAT BRITAIN. Colonial Office.
      Annual report on Gambia. 1946-
Previously issued in its numbered
series, Colonial report--annual
(325.342 G787) no. 1107: 1920-
no. 1893: 1938. Suspended 1940.

      1946-63+   325.342
                             G7871Ga

SA   GAMBIA.
      Blue book. 1822-

      1928, 1932-38   J741
                                   G3R2

SB   GAMBIA.
      Quarterly statistical summary.
Summary for Quarters ending Mar. 31
and June 30, 1965, consists of two pages
of figures.

Mar. 31 and June 30, 1965

CA   GAMBIA.
      Trade report.
Issued in series of Sessional papers of
the Legislative Council.
      Title varies: 1947, Report on the
trade and shipping of the colony.

      1947, 1954-56, 1961-62+   HF265
                                     G2A3

CB   GAMBIA.
      Monthly return of statistics: imports
and exports for....

1st quarter, 1965+

AR   GAMBIA. Dept. of Agriculture.
      Annual reports. 1, 1924-

      FRI   1927/28, 1928/29,
                1932-36, 1946-47,
                1949-51

GERMANY

see also GERMANY (DEMOCRATIC REPUBLIC,
            1949-)
         GERMANY (FEDERAL REPUBLIC,
            1949-)
         GERMANY (OCCUPATION ZONES)

SA   GERMANY. Statistisches Reichsamt.
      Statistisches Jahrbuch für das Deutsche
Reich.... 1880--1940-41//

1900-38
wanting: 1903, 1907, 1908   314.3
                                         G373

      HL   [1889-1938]   HA1232
                                        G373

      FRI   1938; 1940-41,
                v. 1 and 2

SA GERMANY. Statistisches Reichsamt.
    Statistik des Deutschen Reichs. v. 1-
    63, 1873-83; n.s. v. 1-   , 1884-1939//
    Continued by Germany (Federal
    Republic). Statistisches Bundesamt.
    Statistik der Bundesrepublik Deutschland.

    HL v. 6, pt. 2; 44; 95;
        120; 139; 151; 155;
        200; 201; 212-226;
        228-559: 1882;
        1886; 1900; 1907-08;
        1908-39

SA GERMANY. Reichsministerium des
    Innern.
    Handbuch für das Deutsche Reich ....
    1874-
    Superseded by Handbuch für die
    Bundesrepublik Deutschland. 1953-

    HL 1874, 1876-77, 1879-   JN3204
        1931, 1932, 1936   A3

SA GREAT BRITAIN. Dept. of Overseas
CA   Trade.
    ... Report on the economic and finan-
    cial conditions in Germany.... [1919-36.]
    1919 issued in the series of Parlia-
    mentary Papers as Parliament. Papers
    by command. Cmd. 752.
    Title varies.

    1919, 1920, 1922-36   380.005
                                     G786

    HL 1924, 1936   HC286.3
                                     G786

SA GERMANY. Auswärtiges Amt.
    Nachrichten für Handel und industrie.
    1899-1919//
    Early years issued by Reichsamt des
    Innern.
    Title varies slightly.

    1901-09, 1913[3], 1914-15   330.5
                                     G373

SA **GERMANY.
    Staatshandbuch für Gesetzgebung, Ver-
    waltung und Statistik des Norddeutschen
    Bundes und des deutschen Zollvereins;
    unter Benutzung amtlicher Materialien
    .... 1-2, 1868-69//
    Continued under Germany as Annalen
    des Deutschen Reiches.

SB GERMANY. Statistisches Reichsamt.
    Vierteljahreshefte zur Statistik des
    Deutschen Reichs. v. 1, 1892-1941(?)
    Earlier series formed part of Statistik
    des Deutschen Reichs, 1873-76. This
    series continues in part, Monatshefte zur
    Statistik des Deutschen Reichs.

    1900-41
    wanting: 1915-18, v. 24-   314.3
        27; 1938, pt. 3   G373v

    HL 1910-12, 1914-18,
        1920-22, 1930-32,
        1939, 1942

SB **GERMANY. Statistisches Reichsamt.
    Statistisches Handbuch für das
    Deutsche Reich. 1-2, 1907//

SB GERMANY. Statistisches Reichsamt.
    Monatshefte zur Statistik des Deutschen
    Reichs. 1-16, 1884-1891.
    Earlier series formed part of Statistik
    des Deutschen Reichs, 1877-84. Contin-
    ued by Monatliche Nachweise über den
    auswärtigen Handel ... and Vierteljahres-
    hefte zur Statistik des Deutschen Reichs.

CB GERMANY. Statistisches Reichsamt.
    Monatliche Nachweise über den aus-
    wärtigen Handel Deutschlands. 1892-
    1939//
    Monthly; cumulative yearly figures
    in Dec. issue.
    July 1914-June 1920 and Jan.-Apr.
    1921 not published.
    Continues in part Monatshefte zur Sta-
    tistik des Deutschen Reichs.

    1910-July 1939   382.43
                                     G373m

    HL 1910-June 1914,
        1925-26
        wanting: Dec. 1925

CB **GERMANY. Reichsministerium des
    Innern.
    Berichte über Handel und Industrie.
    v. 1-23, no. 5, Oct. 19, 1899-Nov. 4,
    1918//

AA   GERMANY.
     Ernährungswirtschaftliche Leistungen
der Landesbauernschaften in den Kriegs-
jahren 1939-43. Fassung II. Feb. 1945.

    FRI   1939-43 (1 vol.)

AA   GERMANY. Reichsnährstand.
     Ernährungswirtschaftliche Bilanzen
der Landesbauernschaften.

    FRI   1936/37, 1937/38 (pub-
        lished May 1939)

AA **GERMANY. Reichsministerium für
    Ernährung und Landwirtschaft.
     Nachrichten für Bauern- und Land-
arbeiterräte. 1, 1919/20//

AB **GERMANY. Reichministerium des
    Innern.
     Berichte über Landwirtschaft. no. 1-
41, 1907-19//
     No. 1-4 as Bericht des deutschen
Landwirtschaftsrats .... Continued
under Reichsministerium für Ernährung
und Landwirtschaft.

SA   PRUSSIA. Statistisches Landesamt.
     Statistisches Jahrbuch.

    HL   1914-15, 1922-33   HA1291
                                 A4

SA   PRUSSIA. Statistisches Landesamt.
     Preussische Statistik. (Amtliches
Quellenwerk) .... no. 1, 1861-
     Made up of several series consecu-
tively numbered under this title.

    no. 232, 1912   314.31
                              P972

SA **PRUSSIA. Statistisches Landesamt.
     Statistisches Handbuch für der
Preussischen Staat. 1-4, 1888-1903//

SB   PRUSSIA. Statistisches Landesamt.
     Zeitschrift des preussischen statisti-
schen Landesamt 1860-

    HL   1921-34

GERMANY
(Democratic Republic, 1949–)

SA   GERMANY (DEMOCRATIC REPUBLIC,
     1949-). Staatliche Zentralverwal-
tung für Statistik.
     Statistisches Jahrbuch der Deutschen
Demokratischen Republik. 1955-

    1957, 1963-65+   314.3
                             G381s

    HL   1955-64+   HA1248
                             D4A24

    FRI   1963

SA   GERMANY (DEMOCRATIC REPUBLIC,
     1949-). Staatliche Zentralverwal-
tung für Statistik.
     Statistical pocket book of the German
Democratic Republic. 1959-

    1959-61, 1964-65+   HA1248
                             A2A335

    HL   1959-61   HA1248
                             D4A262

SA   GERMANY (DEMOCRATIC REPUBLIC,
     1949-). Staatliche Zentralverwal-
tung für Statistik.
     East German statistical yearbook.
1955-
     Translation of its Statistisches Jahr-
buch der Deutschen Demokratischen
Republik.

    1955   314.3
                             G381sE

    HL   1955   HA1248
                             D4A252

SB   GERMANY (DEMOCRATIC REPUBLIC,
     1949-). Staatliche Zentralverwal-
tung für Statistik.
     Statistische Praxis. Statistical report.
Monthly.

SB   GERMANY (DEMOCRATIC REPUBLIC,
     1949-). Staatliche Zentralverwaltung für Statistik.
    Vierteljahreshefte zur Statistik der Deutschen Demokratischen Republik. Quarterly.

GERMANY
(Federal Republic, 1949–)

SA   GERMANY (FEDERAL REPUBLIC,
     1949-). Statistisches Bundesamt.
    Statistisches Jahrbuch für die Bundesrepublik Deutschland. 1952-

	1953-65+	HA1232 A45
HL	1952-65+ wanting: 1957, 1958	HA1233 A2
FRI	1952, 1954, 1962	

SA   GERMANY (FEDERAL REPUBLIC,
     1949-). Statistisches Bundesamt.
    Statistisches Taschenbuch für die Bundesrepublik Deutschland. 1953-

    1961   314.3
                                            G7375t

SA   GERMANY (FEDERAL REPUBLIC,
     1949-). Statistisches Bundesamt.
    Handbook of statistics.
    Translation of its Statistisches Taschenbuch.
    Published every 3 years.

	1961, 1964+	HA1232 A463
HL	1958	HA1233 A19
FRI	1961	

SA   GERMANY (FEDERAL REPUBLIC,
     1949-). Statistisches Bundesamt.
    Statistik der Bundesrepublik Deutschland. no. 1, 1950-no. 266, 1961//
    Continued by many Fachserie of the same agency.

    no. 1, 1950-no. 266, 1961   330.943
    wanting: several issues   G3735

SB   GERMANY (FEDERAL REPUBLIC,
     1949-). Statistisches Bundesamt.
    Wirtschaft und Statistik.
    Began publication with Jan. 20, 1921, issue; suspended Oct. 1944-Mar. 1949; resumed publication with new series, v. 1, Apr. 1949.

    1921-39, 1941, 1943-44,   314.3
      May 1949-64+   G373w

    FRI   Current two years only.

CA   GERMANY (FEDERAL REPUBLIC,
     1949-). Statistisches Bundesamt.
    Fachserie G. Riehe 5. Special trade according to the Classification for Statistics and Tariffs (CTS).

CB   GERMANY (FEDERAL REPUBLIC,
     1949-). Statistisches Bundesamt.
    Fachserie G. Reihe 1. Aussenhandel: Zusammenfassende Uebersichten. Reihe 2. Aussenhandel: Spezialhandel nach Waren und Ländern. 1962-
    Supersedes its Aussenhandel der Bundesrepublik Deutschland, Teil I and II.

    Reihe 1, Jan. 1962-65+   HA1231
    Reihe 2, Jan. 1962-65+   G4G
                                              Ser. 1 and 2

CB   GERMANY (FEDERAL REPUBLIC,
     1949-). Statistisches Bundesamt.
    Der Aussenhandel der Bundesrepublik Deutschland. Oct. 1949-Dec. 1961//
    Supersedes Monatliche Aussenhandelsstatistik des Vereinigten Wirtschaftsgebietes, issued by Statistisches Amt des Vereinigten Wirtschaftsgebietes.
    Continued in its Fachserie G, Reihe 1 and 2.

    Teil I, July 1951-Dec.   382.43
      1961; Teil II, July   G376
      1951-Dec. 1961
    wanting: Teil II, Feb. 1957

AA   GERMANY (FEDERAL REPUBLIC,
     1949-). Bundesministerium für Ernährung, Landwirtschaft und Forsten.
    Agrarstatistische Arbeitsunterlagen.

    FRI   1951/52-1965/66+

AA   GERMANY (FEDERAL REPUBLIC, 1949-). Bundesministerium für Ernährung, Landwirtschaft und Forsten.
Statistisches Jahrbuch über Ernährung, Landwirtschaft und Forsten.

FRI   1956-64+

AB   GERMANY (FEDERAL REPUBLIC, 1949-). Bundesministerium für Ernährung, Landwirtschaft und Forsten.
Statistischer Monatsbericht .... Aug. 1950-

FRI   Aug. 1950-May 1966+

GERMANY
(Occupation Zones)

The statistical figures appearing in the two SA entries should be used with great caution and care because of the extenuating circumstances at the time of their compilation.

SA   GERMANY. Office of Military Government for Germany. U.S. Ministerial Collecting Center. Economic Division.
Statistisches Handbuch von Deutschland.

     FRI   Teil I; II; II: Annex 1-3;
             III A-B (Sect. 1-2);
             IV; V. 1946

SA   GERMANY (Territory under Allied Occupation, 1945- U.S. Zone). Council of Land Minister-Presidents.
Statistisches Handbuch von Deutschland, 1928/1944.

1928/44   314.3
                               G374

HL   1928/44   HA1233
                               A3

SB   U.S. Civil Affairs Division. [Army Dept.]
Monthly report of the military governor, Office of Military Government for Germany (U.S.). Oct. 1947-Sept. 1949//
    For earlier see War Dept. W 1.72:
    Includes: Statistical Annex, no. 8, Oct. 1947-no. 29, July 1949.
    Statistical series included in Annex continued in Germany (Federal Republic, 1949-). Statistisches Bundesamt. Wirtschaft und Statistik.

no. 28-50: Oct. 1947-   U.S.
Sept. 1949   M105.7:

SB   GERMANY (Territory under Allied Occupation, 1945- French Zone). Direction Générale de l'Economie et des Finances.
Bulletin statistique.

no. 3-6, 9-11; 1946-49   314.3
                                    G376
                                    f

SB   U.S. War Dept.
Monthly report of the military governor, Office of Military Government for Germany (U.S.). Dec. 1945-Sept. 1947.
    For later (no. 28-50, Oct. 1947-Sept. 1949) see Civil Affairs Division. [Army Dept.]
    Includes: Statistical Annex, no. 1, Mar. 1947 (monthly)-no. 7, Sept. 1947.

no. 5-27: Dec. 1945-   U.S.
Sept. 1947   W 1.72:

SB   GERMANY (Territory under Allied Occupation, 1945-). Control Commission. (British Element)
Monthly statistical bulletin.

v. 2: 11-v. 4: 7, Nov.   330.943
1947-July 1949   C397

AA   GERMANY (Territory under Allied Occupation, 1945-). Bipartite Control Office. Food, Agriculture and Forestry Group. Dept. VI. Statistics.
Statistics of German agriculture.
French Zone by land. Bizone and Trizone.

FRI   1934-49

AA   GERMANY (Territory under Allied Occu-
          pation, 1945-   ). Bipartite Control
          Office and Agricultural Group.
          Food and agriculture statistics. Basic
          statistics on agriculture and food.
          Bizone and French Zone.

      FRI   Dec. 1948

AA   GERMANY (Territory under Allied Occu-
          pation, 1945-   U.S. Zone). Office
          of the Military Government.
          Agricultural statistics.

      FRI   1937-45

AA   GERMANY (Territory under Allied Occu-
          pation, 1945-   ). Bipartite Control
          Office. Food and Agriculture Group.
          Food and agriculture statistics.

      FRI   1935-47

GHANA

see also GOLD COAST (COLONY) before 1957

SA   GHANA. Central Bureau of Statistics.
          Statistical yearbook. 1961-

      1961, 1965+   HA1977
                                 G6A264

      FRI   1961

SA   GHANA. Central Bureau of Statistics.
          Economic survey. 1951(?)-
          1953-55 by Gold Coast (Colony) Minis-
      try of Finance; 1957-60 by Ghana Minis-
      try of Finance.

      1953-64+   330.9667
                                 G621

SB   GHANA. Central Bureau of Statistics.
          Digest of statistics. Apr. 1952-
          Title varies: v. 1-2, no. 1, Apr.
      1952-Feb. 1953, Economic and statistical
      bulletin of the Gold Coast.

      v. 3: no. 4, v. 4-13;   316.67
          1955-Dec. 1964+   G618d
      wanting: v. 7, no. 4

SB   GHANA. Central Bureau of Statistics.
          Statistical newsletter. Monthly.

CA   GHANA. Central Bureau of Statistics.
          Report on external trade of Ghana.
          Issued 1937-54 by Gold Coast (Colony),
      Office of the Government Statistician;
      1955/56-1957 by Ghana, Office of the
      Government Statistician.

      1931, 1933, 1934, 1936,   382.667
          1947-60+   G618
      wanting: 1958   f

CA   GHANA. Ministry of Commerce and
          Industry.
          Handbook of commerce and industry.
      1957-

      1957-60+   382.667
                                 G411

CB   GHANA. Central Bureau of Statistics.
          External trade statistics of Ghana.
      Jan. 1951-
          Jan. 1951-Dec. 1956 issued by Gold
      Coast (Colony), Office of the Government
      Statistican; Jan. 1957-Dec. 1960 by
      Ghana, Office of the Government Statisti-
      cian.

      v. 1, no. 9-v. 15, no. 5; 382.667
          Sept. 1951-May 1965+   G617
      wanting: v. 9, no. 7, 8, 12   f

AA   GHANA. Division of Agriculture.
          Report. 1920-
          Issued 1920-1953/54 by Gold Coast
      (Colony), Dept. of Agriculture; 1954/55-
      1955/56 by Ghana, Dept. of Agriculture.

      1920-1956/57+   630.6667
      wanting: 1929-30, 1950-51 G618
                                 f

GIBRALTAR

SA　GREAT BRITAIN. Colonial Office.
　　　Annual report on Gibraltar. 1946-
　　　Previously issued in its numbered series, Colonial reports--annual (325.342　G787) no. 1081: 1920- no. 1900: 1938. Suspended 1940.

　　1946-63+　　　　　　　　　325.342
　　　　　　　　　　　　　　　G7871Gb

SA　GIBRALTAR.
　　　Blue book. 1821(?)-

　　1918-23, 1946-47　　　　　354.468
　　　　　　　　　　　　　　　G447

CA　GIBRALTAR. Port Dept.
　　　Report. Annual.

GILBERT AND ELLICE ISLANDS

SA　GREAT BRITAIN. Colonial Office.
　　　Annual report on the Gilbert and Ellice Islands. 1948-
　　　Previously issued in its numbered series, Colonial reports--annual (325.342　G787) no. 1088: 1919-20-- no. 1879: 1937. Suspended 1940.

　　1948-61+　　　　　　　　　325.342
　　　　　　　　　　　　　　　G7871Gi

SA　GILBERT AND ELLICE ISLANDS.
　　　Blue book.

GOLD COAST (Colony)

see also GHANA after 1957

SA　GREAT BRITAIN. Colonial Office.
　　　Annual report on Gold Coast. 1946-
　　　Previously issued in its numbered series, Colonial reports--annual (325.342　G787) no. 1082: 1919- no. 1919: 1938/39. Suspended 1940.

　　1946-54　　　　　　　　　　325.342
　　　　　　　　　　　　　　　G7871Go

SA　GOLD COAST (COLONY). Office of the
CA　　Government Statistician.
　　　Statistics of external trade and shipping and aircraft movements. 1935-53.
　　　Its Statistical Abstract no. 1.

　　1935-53 (1 vol.)　　　　　382.667
　　　　　　　　　　　　　　　G618s

SA　GOLD COAST (COLONY).
　　　Blue book. 1846-(?)

　　1914, 1917-19, 1921,　　　J741
　　1922/23, 1924/25-1937　　G6R2

CA　GOLD COAST (COLONY). Commerce Division.
　　　Handbook of trade and commerce. 1951-

　　1953, 1955　　　　　　　　382.667
　　　　　　　　　　　　　　　G619

CA　GOLD COAST (COLONY). Customs and Marine Dept.
　　　Annual report on the customs revenue, trade and shipping of the Colony. 1924-

AB　GOLD COAST (COLONY). Dept. of Agriculture.
　　　Bulletin. no. 1, 1926-38(?)

　　no. 7, 16, 18-20, 22-23,　630.6669
　　26-27, 29-32, 35;　　　　G618b
　　1927-38

GREAT BRITAIN

Complete files of many of the serials listed are to be found in the Sessional papers of Parliament, cited Brit. Docs.

SA　GREAT BRITAIN. Central Statistical Office.
　　　Annual abstract of statistics. no. 1- 1840/53-
　　　Each no. contains statistics for each of 15 preceding years.
　　　Nos. 1-83 issued by the Board of Trade and in conjunction with the Secretary of Labour and the Registrar General

with title, Statistical abstract for the United Kingdom. These numbers included in Parliamentary Papers.
    Publication suspended during the war.

no. 1-82; 1840/53-1924/37	Brit. Docs.
wanting: 1862/76	
no. 84, 1935/46- no. 102, 1965+	314.2 G787an

SA GREAT BRITAIN. Board of Trade.
    Statistical abstract for the principal and other foreign countries in each year from 1860/72-1912//
    No. 1-4 have statistics 1860-72, 1860-1875/76; no. 5-7, 1876/77-1878/79 have statistics from 1865, 1866, 1868 respectively; no. 8-39, 1879/80-1912 have statistics for the ten-year period preceding.

1860/72-1912//	Brit. Docs.

SA GREAT BRITAIN. Board of Trade.
CA   Statistical abstract for the British Empire in each year from 1889 to 1903-

1889/1903-1899/1913	Brit. Docs.
FRI 1899/1913	

SB GREAT BRITAIN. Central Statistical Office.
    Monthly digest of statistics. 1, Jan. 1946-

Jan. 1946-Feb. 1966+	314.2 G787a

CA GREAT BRITAIN. Customs and Excise Dept. Statistical Office.
    Annual statement of trade of the United Kingdom with British countries and foreign countries 1853-

1907, 1921/23, 1927-40, 1944-63+	382.42 G786

CA GREAT BRITAIN. Board of Trade.
    Annual statement of accounts relating to the trade and navigation of the United Kingdom.

CA GREAT BRITAIN. Board of Trade.
    Trade of the United Kingdom with selected countries.
    Discontinued 1955.
    Title varies: 1937-50, Overseas trade of the United Kingdom.

1937, 1948-51, 1953-55 (1948 includes figures for 1938-48)	382.42 G786o

CA GREAT BRITAIN. Board of Trade.
    Accounts relating to the trade of the United Kingdom for the years 1938, 1944, and 1945.
    Previously, accounts for imports and exports published separately.

1938, 1944, and 1945 (1 vol.)	382.42 G786ct

CA GREAT BRITAIN. Board of Trade.
    Accounts relating to the export trade of the United Kingdom for the years ....

1938, 1942, and 1943; 1939, 1940, and 1941; 1942, 1943, and 1944	382.42 G786c

CA GREAT BRITAIN. Board of Trade.
    Accounts relating to the import trade and the re-export trade of the United Kingdom for each of the years 1938-1944.

1938-44	382.42 G786ci

CA GREAT BRITAIN. Board of Trade.
    ... Annual statement of the navigation and shipping of the United Kingdom ....
    1914-16 never published. 1917 gives comparative statistics for 1913-17.

1871-1913, 1917-20	Brit. Docs.
1921-38	382.42 G786a

CA GREAT BRITAIN. Dept. of Overseas Trade.
    [Report] 1920-
    The reports in this series are listed separately by country. The series is classed together, however, 380.005 G786.

CB	GREAT BRITAIN. Board of Trade. Report on overseas trade. v. 1, Feb. 1950-		
	Feb. 1950-Dec. 1965+ wanting: v. 7, no. 3-12	382.42 G786r	

GREAT BRITAIN
(Commonwealth)

CB	GREAT BRITAIN. Board of Trade. Board of Trade journal. Monthly 1886-1909(?); weekly, 1918- Beginning 1965, statistics of shipping movements included. Index: 1-14, 1886-93; 15-20, 1893-96.
	v. 1-67: July 1886-Dec. 1909; Suppl. 1915-17; n.s.: v. 101, 1918- v. 186, 1965+ 380.005 B662

SA	GREAT BRITAIN. Board of Trade. The Commonwealth and the sterling area; statistical abstract. 1850/63- Publication suspended 1939-46. Issued 1854/68-1933/47 in the series of Parliamentary Papers as Papers by command. No. 64, 66, and 68-69 contain trade and commerce section only. Title varies: no. 1-40, Statistical abstract for the several colonies and other possessions of the United Kingdom (other variations in title). Current title begins with issue for 1949/52.
	1851/64-1933/47 Brit. Docs. wanting: 1852/65, 1853/67
	1883/97, 1893/1907, HA1111 1929/38, 1933/47, A3 1950-65+
	FRI 1899/1913, 1909/23, 1936/45, 1949/52

AA	GREAT BRITAIN. Ministry of Agriculture, Fisheries and Food. Agricultural statistics, United Kingdom. Issued in two parts. 1921-39 Summaries of Agricultural Statistics for United Kingdom appear in Pt. I of Agricultural statistics, England and Wales.
	1939/44-1962/63+ 630.642 G786ab
	FRI 1939/44-1961/62

SA	GREAT BRITAIN. Colonial Office. Colonial. no. -
	no. 2-361, 1920-65+ 325.342 wanting: no. 1, 175, G784 282, 290

SA	Commonwealth Economic Committee. Annual report. 1933/35-
	1933/35-1963/64+ 338.1042 G78a

AA	GREAT BRITAIN. Ministry of Agriculture, Fisheries and Food. Agricultural statistics, England and Wales. 1870-1920 issued in Parliamentary Papers. 1921-39, Pt. I, includes summaries for Great Britain and for United Kingdom.
	1870-1920 Brit. Docs.
	1921-1962/63+ 630.642 wanting: v. 57: I, 1922 G786a and v. 58: I, 1923

SA	GREAT BRITAIN. Colonial Office. Digest of colonial statistics. 1952- Published as a companion volume to the Monthly digest of statistics issued by Gt. Brit. Central Statistical Office. Quarterly to July 1962. Annual 1963-
	Mar./Apr. 1952-1964+ 314.2 G7862

AB	GREAT BRITAIN. Ministry of Agriculture and Fisheries. Bulletin.
	3-143, 1931-49 630.642 wanting: many issues G786b

SA	GREAT BRITAIN. Colonial Office. Economic survey of the colonial territories. v. 1-7, 1951. Issued as its Colonial no. 281-1/7.
	1951, v. 1-7 325.342 G784

SA   GREAT BRITAIN. Colonial Office.
       Economic survey of the Colonial
Empire.
       Issued as its Colonial no. 95-

1932, 1933, 1936, 1937   325.342
                                          G784

SA   GREAT BRITAIN. Colonial Office.
       Colonial reports--annual. no. 1,
1889-
       Continues Papers relating to H. M.
colonial possessions, 1851-87; Reports,
H. M. colonial possessions, 1888-91.
       Continued, 1946-   by its unnumbered
Annual report on Aden, etc.
       Suspended 1940.

no. 1-1071, 1889-1935   Brit. Docs.
no. 1072-no. 1919,   325.342
   1936-40   G787

SA   GREAT BRITAIN. Board of Trade.
       Statistical tables relating to British
colonies, possessions, and protectorates
.... 1-37, 1854-1912//
       1-26, 1854-1901 as Statistical tables
relating to the colonial and other possessions of the United Kingdom.

1-37; 1854-1912   Brit. Docs.

CA   Commonwealth Economic Committee.
       Commonwealth trade.

1950-64+   382.42
                                        C734

CA   GREAT BRITAIN. Board of Trade.
       [Reviews of commercial conditions]
1944-
       Issued by the Dept. of Overseas
Trade, 1944-45; by the Export Promo-
tion Dept. (later the Commercial Relations and Exports Dept.) of the Board of
Trade, 1947-48.
       Consists of separate pamphlets on
commercial conditions in many foreign
countries.

1944-53   380.005
                                     G786r

CA   Commonwealth Economic Committee.
       Reports. 1-36, 1925-52.
       No. 1-10 in series of Parliamentary
Papers.

no. 1-10   Brit. Docs.
no. 11-36: 1925-52   338.1042
wanting: 15, 25, 34   G78

CB   GREAT BRITAIN. Board of Trade.
       Overseas trade accounts of the United
Kingdom. Feb. 1948-
       Title varies: 1848-1964, Trade and
navigation. Accounts relating to trade
and navigation of the U.K. for each
month.
       Beginning with 1965, the statistics of
shipping movements are no longer
included, but are published each month
in its Board of Trade journal.

1922-40, Jan. 1945-65+   382.42
wanting: Sept., Dec. 1945, G786t
   Mar. 1946, Nov. 1955,
   July 1959

CB   GREAT BRITAIN. Board of Trade.
       Foreign trade and commerce ....
Accounts relating to the trade and com-
merce of certain foreign countries and
British countries overseas ....
       1900-21 in Parliamentary Papers.

1900-21   Brit. Docs.
Sept., Dec. 1925,   382.42
   1926-Mar. 1939   G786f
wanting: several issues
   for 1927

AR   Commonwealth Agricultural Bureaux.
       Executive Council.
       Report.
       Began publication 1929/30. 1929/30-
1946/47 issued by Council under the
Bureaux' earlier name: Imperial Agricultural Bureaux.

1932/33-1963/64+   630.642
                                      G784

GREECE

SA   GREECE. Ethnikē Statistikē Hypēresia.
     Statistical yearbook of Greece. 1930-
     Publication suspended 1940-53.
     Vols. for 1930-39 called v. 1-10; for 1954- called v. 1- of the "post-war series."
     Vols. for 1930-39 in Greek and French; for 1954- in Greek and English.

1930-32, 1934-38,	HA1351
1954-64+	A28

SA   GREAT BRITAIN. Dept. of Overseas
CA   Trade.
     ... Report on economic conditions in Greece .... [1919-38.]
     1919 issued in the series of Parliamentary Papers as Papers by command. Cmd. 793.
     Title varies.

1919, 1921-24, 1927-28,	380.005
1932, 1933, 1937, 1938	G786

SB   GREECE. Ethnikē Statistikē Hypēresia.
     Statistical bulletin. 1956-
     Supersedes Bulletin mensuel de statistique published by the service under an earlier name: Genikē Statistikē Hypēresia.

1956-65+	314.95
	G793s

CA   GREECE. Ethnikē Statistikē Hypēresia.
     Foreign trade of Greece. 1953-55--

1953-63+	382.495
	G793f

CA   GREECE. Ethnikē Statistikē Hypēresia.
     Summary of external trade statistics of Greece. Monthly.
     Beginning 1956, included in its Statistical bulletin.

CA   GREECE. Ethnikē Statistikē Hypēresia.
     Statistique du commerce de la Grèce avec les pays étrangers. 1904-
     Vols. for 1904-17 issued by Statistikon Grapheion; 1918-35 by Genikē Statistikē Hypēresia.
     Title varies: 1904-20, Statistique du commerce spécial de la Grèce avec les pays étrangers.

1911-15, 1918-19: v. 1-2,	382.495
1920: v. 1-2, 1921,	G793s
1923/25, 1928: v. 1-2,	
1939: v. 1, 1940: v. 1	
wanting: 1921, pp. 1-96	

CB   GREECE. Ethnikē Statistikē Hypēresia.
     Bulletin mensuel du commerce spécial de la Grèce avec les pays étrangers.
     Issued by the Statistique Générale de la Grèce (formerly Direction de la Statistique) of the Ministère de l'Economie Nationale.

Jan./Feb. 1919-1924,	382.495
1932, Dec. 1933, Dec.	G793b
1935, May/July 1939-65+	
wanting: many issues for	
1919-48	

CB   GREECE. Hypourgeion tōn Oikonomikon.
     Bulletin trimestriel du commerce spécial de la Grèce avec les pays étrangers ....
     Continues its Bulletin mensuel ..., 1888-98.

1909: 1, 3-4; 1910: 1, 4;	382.495
1916: 1-4; 1917: 1, 2	G793

CB **GREECE. Hypourgeion tōn Oikonomikon.
     ... Bulletin mensuel du commerce spécial de la Grèce avec les pays étrangers 1888-98//
     Continued as Bulletin trimestriel.

AA   GREECE. Ethnikē Statistikē Hypēresia.
     Agricultural statistics of Greece.

FRI   v. 1, 1961; v. 2, 1962

AA GREECE. Ministère d'Agriculture.
    Direction Générale d'Agriculture.
    Production agricole de la Grèce ....

    FRI 1955/56, 1957/58

AA GREECE. Hypourgeion Ethnikēs
    Oikonomias.
    Statistique annuelle du rendement
    agricole....
    1930- has title Statistique annuelle
    agricole d'élevage.

    1914, 1918, 1920,   630.6495
       1930, 1937   G793s

    HL 1918, 1920, 1937

GREENLAND

SA DENMARK. Grønlandsdepartementet.
AA   Beretninger vedrørende Grønland.
      Issues for 1951-   include Grønlands
      landsraads forhandlinger.
      Issued in parts for each year.

    1951-65+   354.981
    wanting: several issues   D3955b

SA DENMARK. Grønlands Styrelse.
    Summary of statistical information
    regarding Greenland. 1942-47.

    1942-47 (1 vol.)   319.8
                                    D397

SA DENMARK. Grønlands Styrelse.
    Beretninger og kundgørelser ved-
    rørende kolonierne i Grønland for aarene
    1909-35.

    1909-12--1933-35   354.981
                                    D3951b

SA DENMARK. Direktoratet for den Konge-
    lige Grønlandske Handel.
    Meddelelser fra Direktoratet for den
    ... for aarene 1882-86--

    1882-86--1907-08   354.981
                                    D395m

AA GREENLAND, SOUTH. Landsraad.
    De grønlandske landsraads forhand-
    linger. 1938-50.
    Each report issued in two sections:
    Sydgrønlands landsraads forhandlinger
    and Nordgrønlands forhandlinger.
    Continued in Beretninger vedrørende
    Grønland. (Denmark. Grønlands-
    departementet.)
    None published for 1942, 1944.

    1938-50   354.981
                                    D396g

AR GREENLAND. Landslaegen.
    Aarsberetning.

GRENADA, WEST INDIES

SA GREAT BRITAIN. Colonial Office.
    Annual report on Grenada. 1948-49--
    Previously issued in its numbered
    series, Colonial reports--annual
    (325.342 G787) no. 1078: 1920-no. 1923:
    1937 and 1938. Suspended 1940-46.

    1948-58, 1961/62+   325.342
                                    G7871Gr

SA GRENADA, WEST INDIES.
    Blue book. 1879-

    1914-18, 1920   354.7298
                                    G826b
                                    f

SB GRENADA, WEST INDIES. Statistical
    Unit.
    Bulletin. Quarterly.

    no. 1, 1963-65+   HA869
                                    G7A3
                                    f

CA GRENADA, WEST INDIES.
    Overseas trade report.

    1921-22, 1936, 1938,   382.7298
       1949-62+   G826

```
CB    GRENADA, WEST INDIES.
          Overseas trade report.  Quarterly.

      Jan./Mar. 1959-              HF157
        Apr./June 1965+            G7A34


AA    WEST INDIES.  Federal Statistical
          Office.
          Agricultural statistics, series 1.
      (1956-58 Survey)

          FRI  no. 4.  The survey
              in Grenada.  1959.


AR **GRENADA.  Agricultural Dept.
          Report.  1911/12+
          Continues its Botanic Station.  Report.
          At head of title: Imperial Dept. of
      Agriculture for the West Indies.


AR **GRENADA.  Botanic Station.
          Report.  1896-1910/11//
          At head of title: Imperial Dept. of
      Agriculture for the West Indies.
          Continued as Agricultural Dept.
      Report.
```

GUADELOUPE

```
SA    FRANCE.  Institut National de la Statis-
          tique et des Etudes Economiques.
          Annuaire statistique de la Guadeloupe.

      1949-53, 1953-57,             317.297
          1959-61                   F815


SA   **GUADELOUPE.
          Annuaire de la Guadeloupe et dépen-
      dances.  1854-
          Continues Almanach de la Guadeloupe
      et dépendances.


CA    GUADELOUPE.  Administration des
          Douanes et Droits Indirects.
          Statistiques du commerce extérieur.
```

GUATEMALA

```
SA    GUATEMALA.  Dirección General de
          Estadística.
          Guatemala en cifras.

      1955-56, 1959-63+             HA811
      wanting: 1961, 1962           A37


SA    GUATEMALA.  Dirección General de
          Estadística.
          Síntesis geográfico-estadística.  1948.

      1948                          317.281
                                    G916
                                    f


SA    GREAT BRITAIN.  Dept. of Overseas
CA        Trade.
          ... Report on economic and commer-
      cial conditions in the republic of Guate-
      mala ....  [1922-37.]
          1924/25, Survey of economic and
      financial conditions in the republics of
      Guatemala, Honduras and El Salvador ....
          1930, Economic conditions in the
      republics of Guatemala, Honduras and
      Nicaragua ....
          Statistics for 1921/22 and 1928 are
      found in the Report on economic and com-
      mercial conditions in the republic of
      Honduras ..., 1921/22, and in Economic
      conditions in the republic of Nicaragua ...
      Report ..., 1928, respectively.

      1922, 1924-25, 1930,          380.005
          1932, 1937                G786


SA    GUATEMALA.  Ministerio de Hacienda.
          Memoria de la Secretaría de hacienda
      y crédito público ....  1880-

      1908, 1914, 1916, 1925-26,    336.7281
          1928-29, 1931, 1933,      G918
          1935-36, 1938-39


SA  **GUATEMALA.  Dirección General de
          Estadística.
          Anuario.  1894, 1898//
          Each year in 3 parts: (1) Territorio y
      población, (2) Criminalidad, etc.,
      (3) Agrícola y forestal, etc.
```

SA **GUATEMALA. Dirección General de
        Estadística.
        Anales estadísticos. 1-2, 1882-83//

SA **GUATEMALA. Dirección General de
        Estadística.
        Informe. 1882-
        Some years as Memoria.

SB   GUATEMALA. Dirección General de
        Estadística.
        Trimestre estadístico. July/Sept.
      1963-

        Jan./Mar.-Oct./Dec.   HA811
          1964+   A4

SB   GUATEMALA. Dirección General de
        Estadística.
        El informador estadístico. Bimonthly.
      Sept. 1962-Feb. 1963//

SB   GUATEMALA. Dirección General de
        Estadística.
        Boletín estadístico. 1946-July/Aug.
      1960//
        Title varies: 1946-55, Boletín; 1956,
      none published (?); 1957, Boletín mensual; 1958-60, Boletín estadístico.

        no. 9, 11-56--v. 13,   317.281
          no. 7-8; 1947-   G916b
          July/Aug. 1960
        wanting: no. 33-34;
          1959, no. 3-6

SB   GUATEMALA. Dirección General de
        Estadística.
        Estadística. 1-138, Sept. 1950-Dec.
      1956//
        Continued by its Boletín mensual.

SB **GUATEMALA. Dirección General de
        Estadística.
        Estadística nacional. Revista mensual. v. 1, no. 1-5, Aug. 1927-
      Jan./Mar. 1928//
        No. 2-3 issued as a double number.

SB **GUATEMALA. Dirección General de
        Estadística.
        Boletín. 1-2, 1913-22//

CA   GUATEMALA. Dirección General de
        Estadística.
        Anuario de comercio exterior.

        1960-62+   HF139
                                   A3

CA   GREAT BRITAIN. Board of Trade.
        Guatemala; economic and commercial
      conditions in Guatemala.

        1951, 1956   380.005
                                   G786ag

AA   GUATEMALA. Dirección General de
        Estadística.
        [Encuesta agropecuaria]
        Each year publication to have a different title.

        1963 y 1964 Producción   S170
          agropecuaria.   A3

AA   GUATEMALA. Secretaría de Agricultura.
        Memoria 1921-

        1936, 1939-40, 1944-45   630.67281
                                   G918

GUINEA, FRENCH

see also FRENCH WEST AFRICA
        GUINEE (REPUBLIQUE), 1958-

SB   GUINEA, FRENCH. Service de la Statistique Générale et de la Mécanographie.
        Bulletin statistique de la Guinée. 1,
      année; Mar. 1955-

        1 année, 1-6, 1955   HA2117
        2 année, 1-12, 1956   G8A3
        3 année, 1-9, 1957
        4 année, numéro
          spécial 2, 1958
        n. s. 3-14, Dec. 1957-
          Nov. 1958
        Suppl. to n. s. 1, Oct.
          1957, Dec. 1957

GUINEA, PORTUGUESE

SB   GUINEA, PORTUGUESE. Serviços de Estatística.
     Anuário estatístico. Annuaire statistique. 1947-

     1949-51   316.657
                                   G964

     FRI   1950-51

SA   **GUINEA.
     Anuário da província da Guiné 1925//
     At head of title: Portugal. Ministério das Colónias.

SB   GUINEA, PORTUGUESE. Secção de Estatística.
     Boletim trimestral da Secção de Estatística. 1938-   Quarterly.

CA   **GUINEA.
     Estatísticas do comércio e navegação 1919-20//

CA   **GUINEA.
     Resumo estatístico aduaneiro da província da Guiné 1910-17//

CB   GUINEA, PORTUGUESE.
     Boletim oficial da colónia da Guiné.
     Includes statistics of foreign trade and public finance.

GUINEA, SPANISH

SA   GUINEA, SPANISH.
     Resúmenes estadísticos del Gobierno General de los Territorios Españoles del Golfo de Guinea.
     1954-55--   issued in series Publicaciónes de la Dirección General de Plazas y Províncias Africanas y del Instituto de Estudios Africanos (Consejo Superior de Investigaciónes Científicas).

     Title varies: Negociado de Estadística; resúmenes del año.

     1941-1946/47, 1950/51,   HA2237
       1954-55, 1956-57   G8A4
                                      f

AA   GUINEA, SPANISH. Dirección General de Marruecos y Colónias.
     Anuario agricola de los territorios españoles del Golfo de Guinea.

     FRI   1942, 1943:1,
               1944, 1947

AA   GUINEA, SPANISH. Dirección de Agricultura.
     Anuario de estadística y catastro.

     1947   630.96718
                                       G964

AB   GUINEA, SPANISH. Servicio Agronomico.
     Ager. v. 1, no. 1, Jan./Mar. 1951-
     Issuing agency varies.

     FRI   v. 1, no. 1-v. 16,
              no. 54/55;
              Jan./Mar. 1951-
              1st trimestre 1965+
              wanting: no. 7-17,
              52-53

AB   GUINEA, SPANISH. Dirección de Agricultura de los Territorios Españoles del Golfo de Guinea.
     Boletín de estadística y catastro de la Dirección de Agricultura. 1944, 1947-

GUINEE (République)

see also GUINEA, FRENCH before Sept. 1958

SB   GUINEE. Direction de la Statistique Générale et de la Mécanographie.
     Bulletin spécial de statistique; statistique et économie.

CA   GUINEE. Service de la Statistique Générale et de la Mécanographie.
     Statistiques du commerce.

H

HAITI (Republic)

SA   HAITI (REPUBLIC). Institut Haïtien de Statistique.
Guide économique de la République d'Haïti. 1964.

       1964   330.97294
                                         H154
                                         t

SB   HAITI (REPUBLIC). Institut Haïtien de Statistique.
Bulletin trimestriel de statistique. July 1951-

       no. 1-44, July 1951-   317.294
         Dec. 1961+   H154
       wanting: no. 25-26, 30-32

CA   GREAT BRITAIN. Board of Trade.
Hayti; economic and commercial conditions in Hayti.
Issued by its Commercial Relations and Exports Dept., in the Overseas economic surveys series.

       July 1952, Aug. 1956   380.005
                                     G786aha

CA **HAITI (REPUBLIC). Customs Administration.
Annual report ... financial adviser-general receiver. 1, 1916/17-
1-7, 1916/17-1922/23 as Report ... Haitian customs receivership.

CB **HAITI. Customs Administration.
Bulletin mensuel. 1, 1924-

AR   HAITI (REPUBLIC). Département de l'Agriculture et de l'Enseignement Professionnel. Service Technique.
Bulletin, no. -
No. 3, 8 are its Rapport annuel, 1924/25, 1925/26.

       no. 3, 8, 17, 27, 31, 33;   630.67294
       1924/25-1943   H154

HAUTE VOLTA

see UPPER VOLTA

HONDURAS

SA   HONDURAS. Dirección General de Estadística y Censos.
Anuario estadístico. 1952-

       1952-59, 1961, 1963-64+   317.283
                                             H771e

SA   HONDURAS. Dirección General de Estadística y Censos.
Honduras en cifras.

       1964+   HA821
                                             A44
                                             f

SA   HONDURAS. Secretaría de Economía y Hacienda.
Informe.

       1919/20, 1925/26,   336.7283
       1934/35-1937/38,   H771m
       1948/49, 1950/51-
       1951/52, 1954/55,
       1961-62+

SA   HONDURAS. Dirección General de Estadística y Censos.
Informe.

       1954/55+   317.283
                                          H881i

SA   GREAT BRITAIN. Dept. of Overseas
CA   Trade.
... Report on economic and commercial conditions in the republic of Honduras .... [1921-38.]
    1921/22, Survey of economic and financial conditions in the republics of Honduras, Nicaragua, El Salvador and Guatemala ....
    1925/27, Report on the economic and financial conditions of the republic of Honduras ... and of the republic of El Salvador ....

                                            (continued)

SA (continued)
CA
    Statistics for 1924/25, 1930, 1932 are found in the <u>Report on economic and commercial conditions in the republic of Guatemala</u>, 1924/25, 1930, 1932, respectively.

    1921-22, 1927, 1928,   380.005
        1934, 1938   G786

SA **HONDURAS. Dirección General de Estadística.
    <u>Anuario estadístico</u>. 1, 1889//

SB HONDURAS. Dirección General de Estadística y Censos.
    <u>Boletín estadístico</u>. 1959(?)-June 1962//

CA HONDURAS. Dirección General de Estadística y Censos.
    <u>Comercio exterior de Honduras</u>. 1953-
    1953-60 issued by Dirección General de Rentas, Centralización de Cuentas y Estadística.

    1953, 1961, 1962, suppl.,   382.7283
        1964: 1-2+   H771

CA HONDURAS. Dirección General de Estadística y Censos.
    <u>Investigación comercial</u>. 1954-57.

    1954-57   382.7283
                                H769

CB HONDURAS. Dirección General de Estadística y Censos.
    <u>Comercio exterior</u>. Quarterly. 1961(?)-
    Each issue cumulative from Jan. of each year.

AA HONDURAS. Dirección General de Estadística y Censos.
    <u>Información agropecuaria; precios recibidos por el productor e índices de precios</u>. no. [1]- 1940-58--
    Subtitle varies slightly.

    1940-58 (1 vol.)   HD361
                                 A4
                                 f

AR HONDURAS. Secretaría de Fomento, Agricultura y Trabajo.
    <u>Informe</u>.
    1886/88-1930/31 have title: <u>Memoria</u>.

    1914/15, 1916/17,   330.97283
    1918/19, 1924/25-   H773
    1928/29, 1937/38,
    1943/44, 1944/45,
    1945/46, 1950/51-
    1952/53

HONG KONG

SA HONGKONG.
    <u>Report</u>. 1946-
    Issued also in the series Colonial annual reports. (Gt. Brit. Colonial Office).
    Preceded by <u>Annual report on the social and economic progress of the people of Hong Kong</u>, issued by Gt. Brit. Colonial Office, in its numbered series, <u>Colonial reports--annual</u> (325.342 G787) no. 1108: 1920-no. 1914: 1938. Suspended in 1940.

    1946-65+   325.342
                                 G7871Ho

SA HONGKONG.
    <u>Blue book</u>. 1844-

    1914-38   J613
    wanting: 1919   R2

SA HONGKONG.
    <u>Historical and statistical abstract of the Colony of Hong Kong</u>. 1841-1930.

    1841-1930 (1 vol.)   DS796
                                 H7A52

SB HONGKONG. Dept. of Statistics.
    <u>Statistical bulletin; monthly supplement to Hongkong Government Gazette</u>.

    Oct. 1960, Dec. 1965-
      Mar. 1966+

CA   HONGKONG. Dept. of Commerce and
Industry.
Review of overseas trade.

1964+   HF240
H6A4

CA   HONGKONG. Dept. of Commerce and
Industry.
Supplement to Hong Kong trade statistics: summary tables.

1964+   HF240
H6A32

CA   HONGKONG. Imports and Exports Dept.
Hong Kong trade returns.
Cover title, 19   Hong Kong trade and shipping returns.

1934, 1937-38   HF240
H6A34

CB   HONGKONG. Dept. of Commerce and
CA   Industry.
Hong Kong trade statistics. Exports.
Monthly; annual reports in Dec. issue.
Title varies: -Dec. 1953, Hong Kong trade returns.

Jan. 1953-Feb. 1966+   382.512
H772te

CB   HONGKONG. Dept. of Commerce and
CA   Industry.
Hong Kong trade statistics. Imports.
Monthly; annual reports in Dec. issue.
Title varies: -Dec. 1953, Hong Kong trade returns.

Jan. 1953-Feb. 1966+   382.512
H772ti

CB   HONGKONG. Dept. of Commerce and
Industry.
Trade bulletin. 1953-
1953-   has title: Trade enquiries bulletin.

Dec. 1957-Dec. 1964+   HF240
H6A36

CB   HONGKONG. Dept. of Imports and
Exports. Statistical Office.
Trade statistics and detailed trade returns.
Monthly, mimeographed.

CB   HONGKONG. Dept. of Commerce and
Industry.
Monthly statistical report.

CB   HONGKONG. Dept. of Commerce and
Industry.
Quarterly report.

AR   HONGKONG. Dept. of Agriculture and
Forestry.
Annual departmental report ....
1950/51-
Supersedes the Reports of the Agricultural Dept., the Fisheries Dept., and the Forestry Dept.

1950/51-1963/64+   630.6512
wanting: 1954/55   H772

AR   HONGKONG. Agricultural Dept.
Report. 1946/47-1949/50(?)//
Superseded by the Annual departmental report of the Director of Agriculture, Fisheries and Forestry, 1950/51-

1946/47-1948/49   630.6512
H771

HUNGARY

SA   HUNGARY. Központi Statisztikai Hivatal.
Statisztikai évkönyv. 1872-
Title varies: 1872-92, Magyar statisztikai évkönyv; 1893-1942, Magyar statisztikai évkönyv. Uj folyam.

1915, 1923-25, 1934,   HA1201
    1938-42, 1949/55-1964+   A52
wanting: 1958

SA HUNGARY. Központi Statisztikai Hivatal.
    Statistical yearbook.
    Issued at irregular intervals, usually
    covering preceding 5-year period.

    1957, 1963   HA1201
                              A523

SA HUNGARY. Központi Statisztikai Hivatal.
    Statistical pocket book of Hungary.
    1958(?)-
    At head of title, 1961-   Hungarian
    Central Statistical Office.

    1958-65+   HA1201
                              A4

    HL   1961-63   HA1201
                              A42

    FRI   1964-65+

SA HUNGARY. Központi Statisztikai Hivatal.
    Teruleti statisztikai zsebkönyv.
    Statistical pocketbook according to areas.

SA HUNGARY. Központi Statisztikai Hivatal.
    Magyar statisztikai zsebkönyv. 1-
    évfolyam; 1933-

    1933, 1956   HA1201
                              A75

SA HUNGARY. Központi Statisztikai Hivatal.
    Magyar statisztikai közlemények. Uj
    sorozat. 1902-1939(?)//
    Issued in Hungarian, Hungarian-
    French, and Hungarian-German.
    Continues Magyar statisztikai köz-
    lemények. Uj folyam.

    no. 1-113, 1902-39   314.39
    wanting: several issues   H937p

    HL   no. 42-115, 1910-39
          wanting: several
          issues

SA HUNGARY. Központi Statisztikai Hivatal.
    Annuaire statistique hongrois. Nou-
    veau cours. 1901-
    Published also in Hungarian and Ger-
    man.

    1913, 1914   314.39
                              H9362

    HL   1913-41

SA GREAT BRITAIN. Dept. of Overseas
CA Trade.
    ... Report on economic and commer-
    cial conditions in Hungary .... [1921-39.]
    Title varies.

    1921-37, 1939   380.005
                              G786

SA **HUNGARY. Központi Statisztikai Hivatal.
    Magyar statiszikai közlemények. Uj
    folyam 1-29, 1893-1901//
    Published in Hungarian and German.
    Continued as Magyar statisztikai köz-
    lemények. Uj sorozat.

SA HUNGARY. Központi Statisztikai Hivatal.
    Ungarisches statistisches Jahrbuch.
    Neue Folge. 1893-1916/18//
    Formerly published in Hungarian-
    German as Magyar statisztikai évkönyv
    ... Stat. Jahrbuch für Ungarn.

    1894, 1915   314.39
                              H9361

SA **HUNGARY.
    Magyar királyi dormany ... évi
    müködéséröl és az orszag közállapotairól
    szóló jelentés és statisztikai évkönyv.
    Report and annual statistics of the activi-
    ties of the government. 1, 1898-
    Contains reports and statistics in con-
    densed form from each ministry and
    department.

SB HUNGARY. Központi Statisztikai Hivatal.
 Statisztikai havi közlemények. 1,
1887-
 Title varies: 1937-56, Statisztikai
negyedévi közlemények.
 Title and text in Hungarian, French,
and German, 1924-41.

July 1914-Sept. 1918,   HA1201
 1933-40, 1941:2,   A73
 1956:1-2, 1957-60,
 Jan. 1965-Dec. 1965+
wanting: Feb. 1915,
 Oct./Dec. 1936, Nov.
 1957, Dec. 1958

SB HUNGARY. Központi Statisztikai Hivatal.
 Statisztikai szemle. 1- évfolyam;
1923-
 Title varies: 1923- Magyar statisztikai szemle.
 Most issues have also title on p. 2 or
4 of cover: Revue hongroise de statistique.

1923-24, 1928-31, 1933-34,   HA1
 1936-Jan. 1966+   H83

SB HUNGARY. Központi Statisztikai Hivatal.
 Statisztikai idöszaki közlemények. 1,
1955-
 Series of statistical bulletins issued
at irregular intervals.

FRI 1-75; 1955-65+
  wanting: several issues

SB HUNGARY. Központi Statisztikai Hivatal.
 Statisztikai tájékoztató. 1950-54.
 No more published.

1950-54   HA1201
wanting: 1950: 3-4   A74

CA HUNGARY. Központi Statisztikai Hivatal.
 Külkereskedelmi forgalma. Commerce extérieur de la Hongrie.
 Issued in its series Magyar statisztikai közlemények. See SA.

AA HUNGARY. Központi Statisztikai Hivatal.
 Mezögazdasági statisztikai zsebkönyv.
 Statistical pocketbook of agriculture.

FRI 1959-60, 1963-65+

AA HUNGARY. Központi Statisztikai Hivatal.
 Mezögazdaságunk a szocialista atszer-
vezes idejem, 1958-1962. Agriculture at
the time of socialist reorganization.
 Issued as its Statisztikai idöszaki közlemények, no. 57, 1962.

FRI no. 57, 1962

AA HUNGARY. Központi Statisztikai Hivatal.
 A mezögazdaság eredményei ....
Agricultural results in ....
 Issued as its Statisztikai idöszaki köz-
lemények, no. 26, 38, -

FRI no. 26, 1958;
   no. 38, 1959

AB HUNGARY. Központi Statisztikai Hivatal.
 Gazdaságstatisztikai tájékoztató.
Dec. 1946-July 1949.

FRI Dec. 1946-July 1949

I

ICELAND

SA ICELAND. Hagstofa.
 Hagskýslur Islands. Statistics of Iceland.
 Series of statistical publications
including census material and volumes
of annual statistics on various subjects.
Applicable series are listed separately.

SA GREAT BRITAIN. Dept. of Overseas
CA   Trade.
 Report on economic and commercial
conditions in Iceland .... 1937.

1937   380.005
                             G786

SA ICELAND. Hagstofa.
 Arbok hagstofu Islands, 1930. Annu-
aire statistique de l'Islande, 1930.
 No more published.

SB   ICELAND. Hagstofa.
      Statistical bulletin. June 1932-
      Issued monthly through 1962, quarterly 1963-

SB   ICELAND. Hagstofa.
      Hagtidindi. Statistical journal. 1916-

CA   ICELAND. Hagstofa.
      Verzlunarskýslur. External trade. 1912-
      Issued as a numbered publication of its Hagskýrslur Islands. Statistics of Iceland.

      1932, 1950-52, 1954-63+   382.491
                                              I15

AA   ICELAND. Hagstofa.
      Búnadarskýslur. Agricultural production statistics. 1912-
      English title varies.
      Issued as a numbered publication of its Hagskýrslur Islands. Statistics of Iceland.

      1949/50--1961-63+   630.6491
                                             I15

INDIA

SA   INDIA (REPUBLIC). Central Statistical Organization.
      Statistical handbook of the Indian Union.
      Title varies: 1962-   Statistical pocketbook.

      1948-64+   315.4
      wanting: 1959   I3922

SA   INDIA (REPUBLIC). Central Statistical Organization.
      Statistical abstract, India. n.s. 1, 1949-
      Supersedes India. Dept. of Commercial Intelligence and Statistics. Statistical abstract for British India, issued 1911/12-1920/21--1946/47.

      1950-1962/63+   HA1713
                                         A732

SA   INDIA. Dept. of Commercial Intelligence and Statistics.
      Statistical abstract for British India, with statistics, where available, relating to certain Indian states. 1911-12/1920-21--1946/47//
      Supersedes Statistics of British India, issued by the Dept. of Statistics, India; and Statistical abstract relating to British India, issued by India Office, Gt. Brit.
      Title and issuing agency vary.
      Superseded by India (Republic). Office of the Economic Adviser. Statistical abstract, India.

      1918-19/1927-28--   315.4
      1930-31/1939-40,   I392a
      1946/47
      wanting: several issues
      2d copies and those issues not found in stacks may be found in Brit. Docs.

SA   INDIA. Dept. of Statistics.
CA   ... Statistics of British India .... 1906/07-1919/20.
      Issued in several parts each year.
      Continuation of "Financial and commercial statistics of British India, 1894-1907" and "Judicial and administrative statistics for British India, Calcutta, 1897-1907."
      Continued by Statistical abstract for British India ....

      1906/07-1919/20   315.4
      wanting: several parts I392

SA **INDIA. Statistical Dept.
CA   Financial and commercial statistics of British India. 1-13, 1892/93-1905/06//
      Continues Dept. of Finance and Commerce. Statistical Branch. Statistical tables relating to British India.
      Continued as Commercial and Intelligence Dept. Statistics of British India.

SA   GREAT BRITAIN. India Office.
      Statistical abstract relating to British India.

      1840/65, 1857/66--   Brit. Docs.
      1917-18/1926-27

SA **INDIA. Dept. of Finance and Commerce.
    Statistical Branch.
    Statistical tables for British India.
1-17, 1875/76-1891/92//
    1-3 as Miscellaneous statistics ....
    Continued in Statistical Dept. Financial and commercial statistics for British India.

SB INDIA (REPUBLIC). Central Statistical
        Organization.
    Monthly abstract of statistics. Oct.
1948-

    v. 4: 4-v. 19:1, 1951-   315.4
        Jan. 1966+   I3922m
    wanting: v. 5: 4, 7;
        v. 7: 4; v. 9: 6; v. 11: 1

SB INDIA (REPUBLIC). Central Statistical
        Organization.
    Weekly bulletin of statistics. n.s.
v. 1-v. 7: 53, Nov. 6, 1948-Dec. 31,
1955//
    Continued by its Monthly abstract of
statistics. Weekly supplement.

    n.s. v. 1-v. 2: 61, Nov. 6,   315.4
        1948-Feb. 24, 1951   I3922w
    wanting: v. 1:17-23

CA INDIA (REPUBLIC). Dept. of Commercial Intelligence and Statistics.
    Annual statement of the foreign trade
of India. 1866/67-1955/56//
    Title and issuing agency vary.

    1919-30, 1918/19-1932/33,   382.54
        1943/44-1948/52: 1-2,   I38
        1952/53-1955/56
    wanting: several issues

CA INDIA. Office of the Economic Adviser.
    Review of the trade of India. 1869-
70/1873-74--
    Earlier vols. have title: Statement of
trade of British India with British possessions and foreign countries.

    1869-70/1873-74--1918-19   Brit. Docs.
    1919-20--1951/52   380.0954
                                I39

CA GREAT BRITAIN. Dept. of Overseas
        Trade.
    ... Report on the conditions and prospects of British trade in India ....
1919-38.
    Title varies.

    1919-38   380.005
                                   G786

CA GREAT BRITAIN.
    Returns as to trade of India. 1813-58.

    1813-58   Brit. Docs.
                                C1859
                                v. 23

CB INDIA (REPUBLIC). Dept. of Commercial Intelligence and Statistics.
    Statistics of the foreign trade of India.
Jan. 1957-
    Supersedes in part its Accounts relating to the foreign trade and navigation of India.
    Dec. 1959- issued in two parts each month, supplements with some issues.

    Jan. 1957-Dec. 1964+   HF239
                                   A44

CB INDIA. Ministry of Commerce and
        Industry.
    The journal of industry and trade.
Aug. 1951-
    Formed by the union of Foreign market review, I and S bulletin, Indian market review, Indian trade bulletin, and Monthly survey of business conditions in India.

    v. 3-v. 15: 4, 1953-   330.954
        Apr. 1965+   J86

CB INDIA. Ministry of Commerce.
    Indian trade bulletin. v. 1-7, no. 14;
Jan. 1945-July 16, 1951//
    Published Jan. 1945-June 1, 1947, by
India. Dept. of Information and Broadcasting (Feb. 16-June 1, 1947, for India.
Dept. of Commerce); June 16-July 16,
1947, by India. Dept. of Commerce;
Aug. 1, 1947-Jan. 1950 by India (Dominion). Ministry of Commerce; Feb. 1950-
July 16, 1951, by India (Republic). Ministry of Commerce.

(continued)

CB (continued)

    Absorbed <u>Indian news and notes</u>, Jan. 1945.
    Superseded by <u>The journal of industry and trade</u>.

v. 1-7, 1945-51	HF41
wanting: many issues	I295

CB INDIA. Dept. of Commercial Intelligence and Statistics.
    <u>Accounts relating to the sea-borne trade and navigation of British India</u>.
    Issued   -Mar. 1880 by the Dept. of Revenue, Agriculture and Commerce; Apr. 1880-Mar. 1895 by the Finance and Commerce Dept.; Apr. 1895-Nov. 1899 by the Statistical Bureau; Dec. 1899-May 1905 by the Statistical Dept.; June 1905-Feb. 1914, Nov. 1922-Nov. 1924 by the Commercial Intelligence Dept.

1908/09-1926/27, 1950/51	382.54
wanting: several issues	I39

AA INDIA. Ministry of Agriculture. Directorate of Economics and Statistics.
    <u>Indian agricultural statistics</u>.
    Began publication with "11th issue," 1890-91, in continuation of <u>Returns of agricultural statistics of British India and the native state of Mysore</u>.
    Issuing agency varies.
    Title varies: 1890-91--1938/39, <u>Agricultural statistics of India</u>.

1901/02-1905/06, 1908/09,	315.4
1910/11, 1912/13-	I3921a
1937/38, 1942/43-	
1956/57: v. 1-2,	
1960/61: v. 1-	

AA INDIA (REPUBLIC). Directorate of Economics and Statistics.
    <u>Abstract of agricultural statistics</u>.

    FRI 1936/37-1945/46,
          1952-1958/59

AA INDIA (REPUBLIC). Directorate of Economics and Statistics.
    <u>Estimates of area and production of principal crops in India</u>.
    1951/52(?)-   issued in two vols.: v. 1, Summary tables; v. 2, Detailed tables.
    Title and issuing agency vary.

    FRI 1896/97-1954/55+
         wanting: 1925/26,
           1927/28, 1930/31

AA INDIA (REPUBLIC). Directorate of Economics and Statistics.
    <u>Quinquennial average yield per acre of principal crops in India</u>.
    Report for 1921/22 issued by the Commercial Intelligence Dept. Title varies slightly.

    FRI 1921/22, 1947/48,
           1951/52

AB INDIA (REPUBLIC). Ministry of Agriculture. Directorate of Economics and Statistics.
    <u>Agricultural situation in India</u>. 1946-

    FRI v. 3: 1-v. 19: 8, Apr.
         1948-Jan. 1966+
         wanting: v. 4: 8, v. 5: 8

INDIA, PORTUGUESE

see PORTUGUESE INDIA

INDOCHINA

SA FRANCE. Haut Commissariat de France en Indochine.
    <u>Annuaire statistique de l'Indochine</u>.
1913-22--
    Issued through 1936-37 by Indochina, French. Service de la Statistique Générale (1932-33 by the Bureau de la Statistique Générale).
    Issues for 1943-46 prepared by Indochina (Federation). Service de la Statistique Générale.

1913-22, 1923-29--	315.96
1932-33, 1936-37,	I41
1941/42-1947/48	

SA   FRANCE. Haut Commissariat de France
SB   en Indochine.
          Bulletin économique de l'Indochine.
       1- 1898-
          Vols. for 1898-19   issued by various agencies of the government of Indochina.
          Publication suspended 1945, no. 2-1947; no more published after June 1952.

       1928: 8, 1929: 1, 1936,   HC441
          1938: 2-6, 1939: 1-6,   A3
          1939-41; n.s. 1949: 10-
          n.s. 1951: 12

SA   FRANCE. Haut Commissariat de France
          en Indochine.
          Statistique générale de l'Indochine.
          Supplément au Bulletin économique de
       l'Indochine.

SA   INDOCHINA, FRENCH. Direction des
          Services Economiques. Service de la
          Statistique Générale.
          Résumé statistique relatif aux années
       1913 à 1940.
          Its Statistique Générale de l'Indochine.
       Supplement to the Bulletin économique de
       l'Indochine.

       1913-40 (1 vol.)

SA   INDOCHINA, FRENCH. Direction des
          Affaires Economiques et Administratives.
          Annuaire économique de l'Indochine.
          Supersedes Annuaire générale, administratif, commercial et industriel de
       l'Indochine.

       1925, 1926/27   330.9596
                                          A615

SA   INDOCHINA.
          Annuaire général, administratif,
       commercial et industriel de l'Indochine.
          Superseded by Annuaire économique
       de l'Indochine and Annuaire administratif
       de l'Indochine.

       1924-25   354.596
                                          A615

SA   INDOCHINA.
          Annuaire général de l'Indochine française. 1887-
          Continues Annuaire de l'Annam,
       Annuaire de la Cochinchine, Annuaire de
       l'Indochine, Annuaire du Cambodge.
          1890-97, pt. 1. Cochinchine et
       Cambodge; 1887-97, pt. 2. Annam et
       Tonkin.

       1924-25   354.596
                                          A615

SB   FRANCE. Haut Commissariat de France
          en Indochine.
          Bulletin statistique mensuel.
          Separately paged monthly statistical
       supplement to the Bulletin économique de
       l'Indochine.

       Dec. 1949, Jan., Apr.,
          June, Aug., Oct.-Dec.
          1950, Jan. 1951, Jan.-
          Mar. 1952

CA   INDOCHINA, FRENCH. Direction des
          Douanes et Régies.
          Tableau du commerce extérieur de
       l'Indochine.

       1934, 1935, 1939 (incl.   382.596
          1936-38 also)   I41

INDONESIA

SA   INDONESIA. Central Bureau of Statistics.
          Statistical pocket book of Indonesia.
       1957-

       1958-63+   HA1811
                                          A34

SA   INDONESIA. Central Bureau of Statistics.
          Statistical abstracts.   -1956//

       1955, 1956   HA1801
                                          A28

SA DUTCH EAST INDIES. Centraal Kantoor voor de Statistiek.
    Statistical pocket-book of Indonesia.
    Title varies: 1940, Pocket edition of the statistical abstract of the Netherlands Indies.

    1940, 1941   HA1811
                              A322

SB INDONESIA. Central Bureau of Statistics.
    Statistik konjunktur. Monthly survey.
    Title varies: -Oct. 1954, Ichtisar bulanan statistik.

    1951-61+   HA1801
                              A35

    FRI Current two years only.

CA INDONESIA. Central Bureau of Statistics.
    Impor dan ekspor.
    Vols. for 1957- have explanatory remarks in English.
    Issued in parts: pt. 1. Impor dan ekspor menurut djenis barang. Trade by commodity; pt. 2. Impor dan ekspor menurut negeri. Trade by country.

    1955, pt. 1: Statistics for   HF247
    1951-55.   A4953

CB INDONESIA. Central Bureau of Statistics.
    Ekspor; menurut djenis barang.
    Export by commodity.
    Vols. for 1957- have explanatory remarks in English.

    Nov., Dec. 1962   HF247
                              A493

CB INDONESIA. Central Bureau of Statistics.
    Impor; menurut djenis barang.
    Import by commodity.
    Vols. for 1957- have explanatory remarks in English.

    Jan./Mar. 1962,   HF247
    Jan./June 1962   A4954

CB INDONESIA. Ministry of Trade and Industry.
    Exports of Indonesia.

CB INDONESIA. Dept. of Economic Affairs, Agriculture and Fishery.
    Exports of Indonesia.

AA INDONESIA. Central Bureau of Statistics.
    Tanam 2 an perdagangen perkebunan.
    Commercial crops of estates.
    Title and issuing agency vary.

    FRI 1953-61+

AA INDONESIA. Central Bureau of Statistics.
    De landvouweportgewassen. Export crops.

    FRI 1938-40, 1948-52

IONIAN ISLANDS

1814-64, under British protection
1864-   annexed to Greece

SA **IONIAN ISLANDS.
    Blue book. 1821-63//

IRAN

SA IRAN. Dept. of Census and Civil Registration.
    Statistical yearbook of Iran.
    In Persian only.

SA GREAT BRITAIN. Dept. of Overseas
CA Trade.
    ... Economic conditions in Persia ... Report.... [1922-37.]
    1935 and 1937: Report on economic and commercial conditions in Iran.

    1922, 1923, 1925-27,   380.005
    1930, 1935, 1937   G786

SB IRAN. Administration Générale des Statistiques et de l'Etat Civil, Administration de Statistique et du Recensement.
Bulletin de statistique.
No. 159 issued in 1956. No longer published.

CA IRAN. General Dept. of Trade Statistics.
Foreign trade statistics of Iran.
Title and issuing agency vary.

1920/21-1921/22, 1938/39-   382.55
    1963/64+   P466
wanting: 1952/53, 1954/55,
    1956/57, 1958/59-1960/61

AA PERSIA. Ministère de l'Economie Nationale.
Agricultural methods and statistics.
In Persian.

AB PERSIA. Ministère de l'Economie Nationale.
Revue agricole et commerciale. Mar. 1916-Mar. 1919; n.s. no. 1-5, Mar.-Aug. 1925; ser. 3, no. 1, Aug. 23, 1928-

IRAQ

SA IRAQ. Central Bureau of Statistics.
Statistical abstract.
Issued 19 -1958 by the bureau under an earlier name: Principal Bureau of Statistics.

1939, 1946-64+   HA1950
                                   I75A3

SA GREAT BRITAIN. Dept. of Overseas Trade.
CA ... Economic conditions in Iraq ... Report .... 1933-37.

1933-37   380.005
                                   G786

SA GREAT BRITAIN. Colonial Office.
Report by His Britannic Majesty's Government to the Council of the League of Nations on the administration.
1920/22-
Reports for 1920-Mar. 1922 by the High Commissioner for Mesopotamia; Apr. 1922- in Colonial Office, Colonial no. -

1920-22--1932   325.342
                                   G784

CA IRAQ. Central Bureau of Statistics.
Bulletin of foreign trade statistics.
1935/36-
Issued 1935/36-1954 by the Dept. of Customs and Excise (called, 1953-54, Directorate General of Customs and Excise); 1956- by the bureau under its earlier name: Principal Bureau of Statistics.
Title varies: 1935/36-1954, Foreign trade statistics; 1956- Statistical abstract for foreign trade.

1945-63+   382.567
                                   I65

CB IRAQ. Central Bureau of Statistics.
SB Bulletin of statistics.
English and Arabic.
Suspended June 1942-Dec. 1950.

July 1952-May 1953,   315.67
    Jan./Mar. 1959-   I65b
    Oct./Dec. 1964+

CB IRAQ. Central Bureau of Statistics.
Monthly bulletin of foreign trade statistics.

AA IRAQ. Central Bureau of Statistics.
Results of the agricultural and livestock census in Iraq for the year ....

FRI 1958-59

IRELAND

SA   IRELAND (EIRE). Central Statistics Office.
<u>Statistical abstract of Ireland.</u>

    1931, 1933, 1938-40,   HA1141
    1942, 1945-64+   A35

SB   IRELAND (EIRE). Central Statistics
CB   Office.
     <u>Irish statistical bulletin. Feasachán staidrimh na hEireann.</u>
     Title varies: Oct. 1925-Sept. 1937, <u>Irish trade journal</u>; Dec. 1937-Dec. 1963, <u>Irish trade journal and statistical bulletin.</u>
     Oct. 1925-Dec. 1937 issued by the Dept. of Industry and Commerce of the Irish Free State; Mar. 1938-Dec. 1963 by the Dept. of Industry and Commerce of Ireland (Eire); Mar. 1964- by the Central Statistics Office of Ireland (Eire).

    v. 40, no. 1, Mar. 1965-   HA1141
    Oct./Dec. 1965+   A38
                                    f

CA   IRELAND (EIRE). Central Statistics
CB   Office.
     <u>Trade statistics.</u>
     Dec. issue cumulative for the year.

    Jan. 1948-Dec. 1965+   382.415
    wanting: Dec. 1948   I68

CA   IRELAND (EIRE). Central Statistics Office.
     <u>Trade and shipping statistics.</u> 1930-
     Suspended 1939-42.
     Vols. for 1930-36 issued by Irish Free State. Dept. of Industry and Commerce; 1937-48 by Ireland (Eire). Dept. of Industry and Commerce.

    1930-31, 1943-62+   382.415
                               I68a

AA   IRELAND (EIRE). Central Statistics Office.
     <u>Agricultural statistics.</u>

    FRI   1960+

AA   IRISH FREE STATE. Ministry of Industry and Commerce.
     <u>Agricultural statistics. 1847-1926. Report and tables ....</u> 1930.

    1847-1926 (1 vol.)   630.9415
                                 I68m

AA **IRISH FREE STATE. Ministry of Industry and Commerce.
     <u>Agricultural statistics. Acreage under crops and numbers of livestock.</u>
     Title varies.
     1921-23, published by Dept. of Agriculture and Technical Instruction; 1924-25 by Dept. of Lands and Agriculture.

AA   IRELAND. Dept. of Agriculture and Technical Instruction.
     <u>Agricultural statistics, Ireland ... General abstracts showing the acreage under crops, and the numbers and descriptions of livestock in each county and province ....</u> 1852/53-1920//

    1877/78-1920   Brit. Docs.

AA   IRELAND. Dept. of Agriculture and Technical Instruction.
     <u>Agricultural statistics of Ireland with detailed report for the year ....</u>
     1847/48-1917//
     Through 1899 issued by Registrar General.

    1847/48-1917   Brit. Docs.

AA   IRELAND. Dept. of Agriculture and Technical Instruction.
     <u>Agricultural statistics, Ireland. Extent in statute acres, produce of the crops and numbers of livestock....</u>
     1881-1908//
     Through 1899 published by Registrar General.
     1882-1905 as <u>Agricultural statistics, Ireland. Tables showing the extent ....</u>

    1882-1908   Brit. Docs.

AB **IRISH FREE STATE. Dept. of Lands and Agriculture.
     <u>Monthly statistical statement.</u> 1921-Apr. 1923//

AR IRELAND. Dept. of Agriculture and
    Technical Instruction.
    Annual general report of the department. 1-21, 1900/01-1920/21//
    Continued by Irish Free State. Dept. of Lands and Agriculture. Annual general report.

    v. 1-19, 1900/01-1918/19   Brit. Docs.

ISLE OF MAN

see also GREAT BRITAIN

SA **ISLE OF MAN.
    Statistical abstract....   1878/87-1905/15//

ISRAEL

SA ISRAEL. Central Bureau of Statistics
    and Economic Research.
    Statistical abstract of Israel. 1949/50-

    1949/50-1965+   HA1931
                               A35

SB ISRAEL. Central Bureau of Statistics
CB and Economic Research.
    Statistical bulletin of Israel.
    Published in parts: 1. Population and social statistics; 2. Economic statistics; 3. Foreign trade; 4. Statistics of prices; 5. English summary.
    Pts. 1, 2, 4 in Hebrew only, 1962-

    v. 1-v. 17: 1/2, July 1949-   315.69
    Jan./Feb. 1966+   I85

CA ISRAEL. Central Bureau of Statistics.
    Israel's foreign trade.
    Issued in its Special series no. -
    1959-61 in 2 parts; 1962 in 4 parts;
    1963-   in 2 parts.
    None published 1954-56.

    1951-64+   315.69
    wanting: no. 76, 1957   I851

ITALY

SA ITALY. Istituto Centrale di Statistica.
    Annuario statistico italiano. Ser. 1, 1878-

    1878, 1897, 1900, 1904-07   314.5
    2. ser., v. 1-9, 1911-   I88
    1919/25
    3. ser., v. 1-2, 5-7,
    1927-28, 1931-33
    4. ser., v. 1-10, 1934-43
    5. ser., v. 1-16, 1944/48-
    1964+

SA ITALY. Istituto Centrale di Statistica.
    Compendio statistico italiano. 1927-
    Suspended 1943-45. Ser. 1, 1927-42;
    Ser. 2, 1946-
    Title varies: 1927-42, Compendio statistico.

    1930, 1934-42, 1946-65+   HA1363
    wanting: 1936, 1950-51   A3

SA ITALY. Istituto Centrale di Statistica.
    ... Annali di statistica ....
    No. 1-10, 1871-77, have title: Annali del Ministero di Agricoltura, Industria e Commercio ... Statistica.

    1. ser., no. 83, 88, 100,   314.5
    1876-77   I87a
    2. ser., v. 1-12, 1878-80
    4. ser., no. 96-107, 1899-
    1904
    5. ser., v. 1-10, 1912-19
    6. ser., v. 1-37, 1929-36
    7. ser., v. 1-7, 1937-43
    8. ser., v. 1-17, 1947-65+
    wanting: 2 ser., no. 100,
    103; 7 ser., v. 5

SA ITALY. Istituto Centrale di Statistica.
    Italian statistical abstract. 1954(?)-
    63//
    Published also in Italian.

    1954-63   HA1362
    wanting: 1960   A5

SA   ITALY. Istituto Italiano per l'Africa.
     Annuario delle colonie italiane e dei
     paesi vicini. 1, 1926-

     HL   1926-1938/39   JV2201
                               A615

SA   ITALY. Direzione General della Statis-
CA   tica.
     Statistica giudizaria civile e commer-
     ciale. 1873-

     1873-90, 1894-95,   314.5
       1898, pt. 1   I87s

SA **ITALY. Ministero di Agricoltura, Indus-
     tria e Commercio.
     Annali. 1-107, 1870-79//
     Continued in three series: Annali
     dell'industria e del commercio; Direzione
     Generale dell'Agricoltura. Annali di
     agricoltura; and Istituto Centrale della
     Statistica. Annali di statistica.

SA **ITALY. Ministero di Agricoltura, Indus-
     tria e Commercio.
     Annali di agricoltura, industria e
     commercio. 1-2, 1862-64//

SB   ITALY. Istituto Centrale di Statistica.
     Bollettino mensile di statistica ....
     1926-
     Suspended Sept. 1943-Aug. 1945.

     anno 10-15, 1935-40; Sept.-   314.5
       Dec. 1945--Dec. 1964+   I88b

     FRI Current two years
            only.

CA   ITALY. Istituto Centrale di Statistica.
     Statistica del commercio con l'estero.
     Earlier vols. published with title:
     Commercio con l'estero.

     1939, 1946-49: v. 1-2,   382.45
       1952-62+   I85ca

CA   ITALY. Ministero dell'Africa Italiana.
     Ufficio Studi e Propaganda.
     ... Statistica del movimento commer-
     ciale marittimo dell'Eritrea, della
     Somalia italiana, della Tripolitania e
     della Cirenaica e del movimento com-
     merciale carovaniero dell'Eritrea ....
     1921 e 1922-
     1925-26 includes "Movimento com-
     merciale marittimo dell'Oltre Giuba dal
     1° luglio 1925 al 20 giugno 1926."
     1928-29 includes "Movimento della
     navigazione marittima della quattro
     colonie."

     1923-24   382.45
                               I892

     HL   1923-24, 1931-32   HF270
                               I88

CA **ITALY. Ministero di Agricoltura, Indus-
     tria e Commercio.
     Annali del commercio. 1913-
     1920/21//(?)
     Suspended 1917-20.

CA **ITALY. Ministero di Agricoltura, Indus-
     tria e Commercio.
     Annali dell'industria e del commercio.
     no. 1-56, 1870-82; Ser. 2, no. 1-198,
     1879-1914//
     1879-82 as Annali dell'industria, del
     commercio e del credito.

CB   ITALY. Istituto Centrale di Statistica.
     Statistica del commercio con l'estero.
     Mensile.

     June 1946-Dec. 1948,   382.45
       Dec. 1953   I85cm

CB   ITALY. Istituto Centrale di Statistica.
     Statistica del commercio speciale di
     importazione e di esportazione. 1870-
     Issued 1870-1917 by Direzione Gene-
     rale delle Gabelle; 1918-June 1935 by
     Direzione Generale delle Dogane e
     Imposte Indirette.

     Dec. 1917-June 1943   382.45
     wanting: several issues   I89

CB   ITALY. Istituto Nazionale Fascista per
     gli Scambi con l'Estero.
     ... Bollettino di informazioni com-
     merciali. anno 1, 1927-
     Continues Bollettino di notizie com-
     merciali published by Direzione Generale
     del Commercio e della Politica Econo-
     mica.

     v. 1-10, 1927-36   382.45
     wanting: no. 30, 46 of   I87
        v. 10

CB   ITALY. Direzione Generale del Com-
     mercio.
     Bollettino di notizie commerciali.
     Continued by Bollettino di informazioni
     commerciali (Italy. Istituto Nazionale
     Fascista per gli Scambi con l'estero).

     v. 50-52, 1924-26   382.45
     wanting: several issues   I86
        of v. 50

CB   ITALY. Direzione Generale delle Dogane
     e Imposte Indirette.
     Movimento commerciale del regno
     d'Italia. 1851-
     Issues for 1851-58 have only statistics
     for Sardinia; 1859, provinces of northern
     Italy; 1860, the provinces of northern
     Italy and Emilia.
     Issuing agency varies.

     1913: 1-3; 1916: pt. 2,   380.0954
        v. 1; 1917: pt. 2, v. 2,   I88m
        pt. 3; 1918; 1919: 1-2;
        1920: 1-2; 1923

CB   ITALY. Ministero di Agricoltura, Indus-
     tria e Commercio.
     Bollettino di notizie commerciali ....
     1-6, 1878-84; Ser. 2, v. 1-18, 1885-
     1901//
     Continued with same title under
     Direzione Generale del Commercio ....

AA   ITALY. Istituto Centrale di Statistica.
     Annuario di statistica agraria.

     1955-65+   630.645
                                   I88ag

     FRI 1954-65+

AA   ITALY. Istituto Centrale di Statistica.
     Annuario statistico dell'agricoltura
     italiana.

     FRI 1939/42, 1943/46,
          1947/50

AA   ITALY. Direzione Generale dell'Agri-
     coltura.
     Annali di agricoltura, serie seconda.
     no. 1-270, 1878-1921//

     no. 229, 258, 260-264,   630.645
        267, 270; 1901-21   I88a

AA   ITALY. Ufficio di Statistica Agraria.
     Catasto agrario del regno d'Italia ....
     1910-11--

     1913-14 (1 vol. in 2 pts.)   630.645
                                   I88c

AB   ITALY. Istituto Centrale di Statistica.
     ... Bollettino mensile di statistica
     agraria e forestale .... anno 1, 1928-
     Issued at the end of each month as a
     special supplement to the Gazzetta
     ufficiale del regno d'Italia.
     Continues Notizie periodiche di
     statistica agraria.

     1928-39, July 1947-   630.645
        Dec. 1949   I88
     wanting: v. 1, no. 1,   1928-
        and Aug., Sept. 1948

AB   ITALY. Ufficio di Statistica Agraria.
     Notizie periodiche di statistica agraria.
     1910/11-1927//
     Continued by Bollettino mensile di
     statistica agraria e forestale.

     Sept., Nov. 1917,   630.645
        June 1919-26   I88

IVORY COAST

see also FRENCH WEST AFRICA

SA   IVORY COAST. Direction de la Statistique et des Etudes Economiques et Démographiques.
     Situation économique de la Côte d'Ivoire. 1960-

1960-63+	330.96668 I96s

FRI   1960

SB   IVORY COAST. Direction de la Statistique et des Etudes Economiques et Démographiques.
     Bulletin statistique mensuel.
     Title and agency name vary slightly.

1955-61, Jan. 1966+	HA2117
1959-61 have suppl.	I8A36
wanting: 1955: no. 8, 1956: no. 8	

CA   IVORY COAST. Service de la Statistique Générale.
     Statistiques du commerce extérieur: importations, exportations, commerce spécial.
     1963 not published.

1954-62+	382.6668 I96e

CB   IVORY COAST.
     Statistiques douanières de la Colonie de la Côte d'Ivoire.
     Annexe au: Journal officiel de la Guinée française.

HL   Jan.-Sept. 1938
      (1 vol.)

AB   IVORY COAST. Ministère de l'Agriculture et de la Coopération.
     Bulletin de liaison. Bulletin bimestriel de l'agriculture.

FRI   1962-Mar. 1966+

J

JAMAICA

SA   JAMAICA. Dept. of Statistics.
     Abstract of Statistics. Annual.
     Issued 1947-53 by the department under its earlier name: Central Bureau of Statistics.
     Title varies: 1947-57, Digest of statistics.

June 1947-Dec. 1952,	317.292
1953-65+	J27d
wanting: 1954, 1955	

SA   GREAT BRITAIN. Colonial Office.
     Annual report on Jamaica. 1946-
     Previously issued in its numbered series, Colonial reports--annual (325.342   G787) no. 1116: 1920- no. 1896: 1938. Suspended 1940-46.

1946--1960-61+	325.342
wanting: 1957	G7871J

SA   JAMAICA.
     Blue book for the island of Jamaica.
1891/92-1945//

1914/15-1938, 1945	354.7292 J27b

SB   JAMAICA. Dept. of Statistics.
     Quarterly abstract of statistics.
Mar. 1961-

no. 1-17, Mar. 1961- Mar. 1965+	317.292 J27q

CA   JAMAICA. Dept. of Statistics.
     External trade of Jamaica.

1947-63+	382.7292 J27

CB JAMAICA. Dept. of Statistics.
    Bulletin. Dec. 1945-

    no. 1-225, Dec. 1945-   317.292
      Jan. 1966+   J27
    wanting: 14, 123-24, 156,
      159-160, 163

AA JAMAICA. Dept. of Statistics.
    Agricultural statistics.

    FRI 1954-55 (1 vol.)

AB JAMAICA. Dept. of Statistics.
    Digest of agricultural statistics.
    1953-
      Irregular.

    1953   630.67292
                                    J27

AR JAMAICA. Dept. of Science and Agriculture.
    Annual report of the Department of Agriculture.

    1925, 1926, 1946   630.6
                                    J27a

JAPAN

SA JAPAN. Prime Minister's Office.
      Bureau of Statistics.
    Japan statistical yearbook. 1949-
      Supersedes a statistical yearbook which was issued in Japanese only and ceased publication with v. 59, 1941.
      Vol. for 1949 contains retrospective figures.

    1949-64+   HA1832
    wanting: 1950   J36

    HL 1949-64+   J   HA1831
        wanting: 1955, 351   A35
          1956   Jso55

SA JAPAN. Prime Minister's Office.
      Bureau of Statistics.
    Statistical handbook of Japan.

    1958, 1964-65+   HA1831
                                    A4

    FRI 1964-65+

SA JAPAN. Prime Minister's Office.
      Bureau of Statistics.
    Statistical abstract of Japan. 1950-
      Abstract edition of the Japan statistical yearbook issued by the same bureau.

    1950   315.2
                                    J352

SA JAPAN. Cabinet Statistical Bureau.
    Résumé statistique de l'empire du Japon.... 1887-
      Issued by the bureau under variant names.

    1889-1915, 1920,   315.2
      1927-35, 1939-40   J351
    wanting: 1891, 1895

SA GREAT BRITAIN. Dept. of Overseas
CA   Trade.
    Report on the commercial, industrial, and financial situation in Japan....
    [1914-1936.]
      1914-19 issued as Parliament. Papers by command. Cmd. 912.
      Title varies.

    1914-24, 1926, 1928-30,   380.005
      1932-34, 1936   G786

SA JAPAN. Cabinet Statistical Bureau.
    Teikoku tokei nenkwan. Annual statistics of the Japanese Empire. 1882-1941.
      In Japanese.

    HL v. 1-59, 1882-1940   J
        wanting: 1882, 1884,   351
          1909   Jn28

SA **JAPAN. Bureau de la Statistique Générale.
    Statistical account of the conditions of the Empire. 31, 1912//
      Only edition issued in English.

SB   JAPAN. Prime Minister's Office.
       Bureau of Statistics.
       Monthly statistics of Japan. July 1961-
       Continues its Monthly bulletin of statistics.

       1, 1961-Feb. 1966+   315.2
                              J352ms

    FRI   1, 1961-Feb. 1966+

SB   JAPAN. Prime Minister's Office.
       Bureau of Statistics.
       Monthly bulletin of statistics. no. 1-135, Mar. 1948-Mar. 1961.

       Mar. 1950-Dec. 1960   315.2
       wanting: Dec. 1951,   J352mb
         Jan.-May 1952

CA   JAPAN. Ministry of Finance.
       Annual return of the foreign trade of Japan. 1882-
       Continues Bureau of Customs. Return of the foreign commerce and trade.

       1964+   382.52
                                     J35an

    HL   1914-19, wanting   J
            several vols. or   678.91
            parts of vols.   Jo57n
            1950, pt. 1-2--
            1960, pt. 1-3

CA   JAPAN. Ministry of International Industry and Trade.
       Foreign trade of Japan.

       1950, 1954, 1958

CA   JAPAN. Bureau of Customs.
       Returns of the foreign trade ....
       Comparative statistics for each year from 1868.
       Title varies.

    HL   1868-99, 1868-1904

CA **JAPAN. Bureau of Customs.
       Annual report of exports and imports at all ports ....

CA **JAPAN. Dept. of Commerce and Industry.
       Statistics of the Department of Commerce and Industry 1, 1924-

CB   JAPAN. Ministry of Finance.
       ... Monthly return of the foreign trade of Japan ....
       Title varies.

       1917, Dec. 1920, Jan.   382.52
       1925, 1931-Sept. 1940,   J35
       Jan.-Oct. 1952, 1953-
       Dec. 1964.

AA   JAPAN. Ministry of Agriculture and Forestry.
       Statistical yearbook of Ministry of Agriculture and Forestry, Japan.
       Japanese, with Arabic numerals in all tables.

    FRI   v. 23-40, 1946-
            1963/64+
            wanting: v. 30, 1953;
            v. 36, 1959

AA   JAPAN. Ministry of Agriculture and Forestry.
       Abstract of statistics on agriculture, forestry, and fisheries. 1924-
       Continuation, with Statistics of the Department of Commerce and Industry, of The statistics of agriculture, industries and commerce, 1919-23.
       Title varies: 1924-1931/32: The statistical abstract; 1940-59, Statistical abstracts.

       1924-1936/37, 1940-64+   630.652
       wanting: 1925, 1928, 1953 J34s

    FRI   1940-64+
            wanting: 1953

AA   JAPAN. Ministry of Agriculture and Forestry.
       Yearbook. Norinsho nempo.
       Figures in statistical tables are Arabic.

    FRI   1955-63+

AA	JAPAN. Ministry of Agriculture and Forestry. Statistics and Survey Division. Handbook of statistics of agriculture, forestry and fisheries. 1953. No more published.		AR CA	JAPAN. Dept. of Agriculture and Commerce. Statistical report. 1905/06 has title: 21st statistical report of the Department of Agriculture and Commerce, Japan.

1953   630.952
         J364

FRI 1953

1905/06, 1915, 1917   315.2
            J353

JORDAN

see also PALESTINE   SA

AA   JAPAN. Dept. of Agriculture and Commerce.
CA   The statistics of agriculture, industries and commerce. 1919-23.
 Continuation of Abstract of the statistics of the Imperial Japanese Department of State for Agriculture and Commerce.
 Continued by the Statistics of the Department of Commerce and Industry and The statistical abstract of the Ministry of Agriculture and Forestry.

1919-23   315.2
         J36

SA   JORDAN. Dept. of Statistics.
 Statistical year book. 1950-

1952-64+   HA1950
         J6A32

SA   JORDAN. Dept. of Statistics.
 Statistical guide to Jordan. no. 1, 1964-

1964+   HA1950
         J6A4

AA   JAPAN. Dept. of Agriculture and Commerce.
CA   Abstract of the statistics of the Imperial Japanese Department of State for Agriculture and Commerce. 1, 1900-

1900   315.2
         J344

CA   JORDAN. Dept. of Customs, Trade and Industry. Statistics Section.
 Annual report; customs, excise, trade and industry.

K

AB   JAPAN. Ministry of Agriculture and Forestry.
 Agriculture, forestry and fishery statistics, monthly.
 In Japanese with English tables.

FRI no. 6-132, 1952-63;
    1964: 4-12, 1965: 1-3

KENYA

see also EAST AFRICA

 The East Africa Statistical Dept. is responsible for the trade statistics of Kenya, Tanzania, and Uganda.

AB   JAPAN. Ministry of Agriculture and Forestry.
 Monthly bulletin of agricultural statistics and research.

FRI July 1953-1965: 2+

SA KENYA. Economics and Statistics Division.
  Statistical abstract. 1961-
  Supersedes the Statistical abstract issued by the Kenya Unit of the East African Statistical Dept. of the East Africa High Commission.

  1961-65+   HA1977
  wanting: 1964   K4A4

SA GREAT BRITAIN. Colonial Office.
  Report on the Colony and Protectorate of Kenya. 1946-
  Previously issued in its numbered series, Colonial reports--annual (325.342 G787) no. 1073: 1918-19--no. 1920: 1938. Suspended 1940-46.

  1946-62+   325.342
                              G7871Ke

SA East Africa High Commission. East African Statistical Dept. Kenya Unit.
  Statistical abstract [of the Colony and Protectorate of Kenya]. 1955-60//
  Superseded by the Statistical abstract issued by the Economics and Statistics Division of Kenya Colony and Protectorate.
  Earlier annual statistics included in its Quarterly economic and statistical bulletin.

  1955-60   316.762
                              E13

SA KENYA.
  Blue book. 1912/13-
  1912/13-1920/21 issued as East Africa Protectorate. Blue book.

  1914-16, 1945-46   354.676
                              K37b

SB KENYA. Economics and Statistics
CB   Division.
  Kenya statistical digest. v. 1, Sept. 1963-

  v. 1: 1-v. 3: 3, Sept.   HA1977
  1963-June 1965+   K4A3

CA East African Common Services Organization. East African Customs and Excise Dept.
  Annual trade report of Kenya, Uganda, and Tanganyika.
  Issued through 1948 by the Commissioner of Customs, Kenya and Uganda. In Jan. 1949, the Customs Dept. of Kenya, Uganda, and Tanganyika were amalgamated under title East African Customs and Excise Dept.
  Reports for 1950-60 published under the authority of the East Africa High Commission.

  1925-64+   382.676
  wanting: 1926, 1928,   K37
    1932, 1935-36

CA KENYA COLONY AND PROTECTORATE.
  Ministry of Commerce and Industry.
  Commerce and industry in Kenya. 1948-61//
  Title varies: 1953-   Notes on commerce and industry in Kenya.

  1953-61   330.9676
                              K37

CA KENYA COLONY AND PROTECTORATE.
  Customs Dept.
  Annual trade report of Kenya and Uganda. 1907/08-

CB KENYA COLONY AND PROTECTORATE.
  Dept. of Trade and Supplies.
  Kenya trade and supplies bulletin.
  Began publication with the Jan. 1956 issue; ceased with v. 8, no. 5, June 1963.
  Superseded by Kenya statistical digest, published by the Economics and Statistics Division.

  v. 6, no. 1-v. 8, no. 5;   382.676
  1961-63   K38

AA KENYA COLONY AND PROTECTORATE.
  Dept. of Agriculture.
  Agricultural census ... Annual report.

  FRI 1925-34, 1938, 1954-56
       Summary of results, 1938

AA KENYA. Dept. of Agriculture.
    <u>Bulletin.</u> no. 1-13, 1914-22; n.s.
no. 1-28, 1925-27; Ser. 3, no. 1, 1929-
Suspended 1923-24.

    FRI 1925-35
        wanting: many issues

AR KENYA COLONY AND PROTECTORATE.
    Dept. of Agriculture.
    <u>Annual report.</u>

1919/20-1925, 1928/29-	630.6676
1938, 1949-61	K37
wanting: 1930, 1948,	
1953, 1955, 1960	

KOREA

SA KOREA (REPUBLIC). Economic Planning Board.
    <u>Korea statistical yearbook.</u>

| 1961-65+ | HA1851 |
| | A49 |

SA KOREA (REPUBLIC). Economic Planning Board.
    <u>Statistical handbook of Korea.</u>

| 1962 | HA1851 |
| | A53 |

SA U.S. Economic Cooperation Administration. Mission to Korea.
    <u>Republic of Korea statistical summation.</u> no. 1, Jan. 1949-

| no. 9, Sept. 1949 (figures | HA1855 |
| for 1946-49) | U63 |

SA **KOREA.
    <u>Statistical abstract ....</u> 1925(?)-

SB KOREA (REPUBLIC). Economic Planning Board.
    <u>Bulletin of statistics.</u>

| May 1961-Feb. 1966+ | 315.19 |
| wanting: several issues | K84b |

SB KOREA (SOUTH). South Korean Interim Government. National Economic Board.
    <u>South Korea Interim Government Activities, U.S. Army Military Government in Korea.</u> Mar. 1946-
    Title varies.

| HL Mar. 1946- | D802 | D802 |
| Sept./Oct. 1948 | K8U58 | J3A3 |

CA KOREA.
    <u>... Report on the foreign trade ....</u>
1885-
    Published by China Inspectorate General of Customs, 1885- ; by Japan Bureau of Customs, (?)
    Title varies.

| 1885, 1911-12 | 382.519 |
| | K84 |

AA KOREA. Ministry of Agriculture.
    <u>Yearbook of agriculture and forestry statistics.</u>

    FRI 1952-53, 1964+

KOREA (Democratic People's Republic)

No applicable publications found.

L

LAGOS

In 1906 Lagos became a province of Southern Nigeria.

SA **LAGOS.
    <u>Blue book.</u> 1862-1905.

LAOS

see also INDOCHINA

SA LAOS (KINGDOM). Service de la Statistique.
    Annuaire statistique de Laos.
    1949/50-
    Earlier statistics for Laos were included in the Annuaire statistique issued by the Service de la Statistique Générale de l'Indochine.

    1949/50-1951/52,   HA1770.5
    1953/54-1956/57   A64

SB LAOS (KINGDOM). Service de la Statistique.
    Bulletin de statistiques du Laos.
    no. 1, 1951-
    Issued 1951-May 1952 by Ministère de l'Economie Nationale. Name of present agency varies.
    Supersedes Bulletin de renseignements économiques.
    Title varies slightly.

    1955-61: [4]   315.96
                              L298b

CB LAOS (KINGDOM). Bureau de la Statistique.
    Bulletin du commerce extérieur [importations-exportations]. Jan.-Mar. 1950-
    Supersedes its Statistique des importations et des exportations, published by the bureau under a variant from of name, Bureau des Statistiques.

LATVIA

SA LATVIAN S.S.R. Centrala Statistikas Párvalde.
    Latviiskaia SSR v tsifrakh.
    Issued by: TSentralnoe Statisticheskoe Upravlenie of the Latvian SSR.

    HL 1961   HA1448
                              L3L36

SA LATVIAN S.S.R. Valsts Statistiská Párvalde.
    Latvijas PSR tautas saimniecība; statistisko datu krajums.
    Latvian and Russian.

    HL 1957   HA1448
                              L3A3

SA LATVIA. Valsts Statistiská Párvalde.
    ... Latvijas statistiská gada gramata.
    ... Annuaire statistique de la Lettonie.

    HL 1920-26, 1932,
        1937/38

SA GREAT BRITAIN. Dept. of Overseas
CA Trade.
    ... Report on economic and commercial conditions in Latvia .... 1925-38.
    Title varies.

    1925, 1927, 1929, 1932,   380.005
    1935, 1938   F786

    HL 1927, 1929

SA LATVIA. Valsts Statistiská Párvalde.
    Bureau de Statistique de L'état.
    Latvijas statistikas atlass. Atlas statistique de la Lettonie.

    HL 1938   HC337
                              L3A2

SB LATVIA. Valsts Statistiská Párvalde.
    Menesa bileten s; statistikai un konjunkturai. Monthly statistical and economic bulletin. 1, Sept. 1926-

    HL Jan. 1934-Oct. 1940

CA LATVIA. Valsts Statistiská Párvalde.
    Latvijas áréja tirdzniecība un kugu kustiba .... Commerce extérieur et mouvement de la navigation de la Lettonie ....

    HL 1920-26, 1929, 1934-35

AA   LATVIA. Valsts Statistiská Párvalde.
<u>Latvijas lauksaimniecîba ....</u>   <u>Etat
de l'agriculture.</u>   1920/25-
1920/25 and 1926/27 in one vol. each.

HL   1920/25

LEBANON

SA   Conseil Supérieur des Intérêts communs
(Syria and Lebanon). Service des
Etudes Economiques et Statistiques.
<u>Recueil de statistiques de la Syrie et
du Liban.</u>   1942-

    HL   1942/43-1944,   HA1941
          1945-47, 1948-49   A3

SA   LEBANON. Service de la Statistique
Générale.
<u>Recueil des statistiques générales.</u>
1946-

SA   GREAT BRITAIN. Dept. of Overseas
CA   Trade.
    ... <u>Economic conditions in Syria.</u>
[1920-38.]
    Title varies: 1936/38-   <u>Report on
economic and commercial conditions in
Syria and the Lebanon.</u>

    1920, 1922, 1923, 1925,   390.005
       1928, 1930, 1932,   G786
       1934, 1936/38

SB   LEBANON. Direction Centrale de la
Statistique.
<u>Bulletin statistique mensuel.</u>   v. 1,
June 1963-
Supersedes <u>Bulletin statistique tri-
mestriel</u> formerly issued by Service de
Statistique Générale.

    June 1963-Dec. 1965+   HA1950
                                   L4A3

SB   LEBANON. Service de Statistique
CB   Générale.
<u>Bulletin statistique trimestriel.</u>
v. 1-12, 1950-62.
Title and text also in Arabic.
Superseded by <u>Bulletin statistique
mensuel</u> issued by Direction Centrale de
la Statistique.

    v. 1-12, 1950-62   315.69
                                    L441

SB   LEBANON. Service de Statistique
Générale.
<u>Bulletin mensuel.</u>
Suspended publication with Dec. 1959.

    May 1950-Dec. 1959   315.69
    wanting: 1953-57   L441bm

SB   **LEBANON. Office Commercial pour la
Syrie et le Liban.
<u>Bulletin économique.</u>   1, 1921-

CA   LEBANON. Direction Centrale de la
Statistique.
<u>Statistiques du commerce extérieur
du Liban.</u>   v. 1-   1960-63--
English and Arabic.

    1960-63, 1961-64+   HF259
                               L4A24

CA   LEBANON. Direction Générale des
Douanes.
<u>Statistiques du commerce extérieur.</u>

    1951-55, 1956-58,   382.569
       1960-63+   L441s

CA   Conseil Supérieur des Intérêts Communs
(Syria and Lebanon). Administration
des Douanes.
<u>Statistiques du commerce extérieur.</u>

    1939/43: v. 1-2, 1944-47   HF259
                                         S95A3

LEEWARD ISLANDS

SA   GREAT BRITAIN. Colonial Office.
      Report on the Leeward Islands.
   1947-54.
      Previously issued in its numbered series, Colonial reports--annual (325.342   G787) no. 1074: 1919-20-- no. 1934: 1938. Suspended 1940-46.

   1947-50, 1953/54   325.342
                                  G7871Le

SA   LEEWARD ISLANDS.
      Blue book. 1888-
      Continues the Blue books of Antigua and Montserrat.

   1922-25   354.7297
                                  L487b

LIBERIA

SA   LIBERIA. Information Service.
      Statistical survey of Liberia.

CA   LIBERIA. Information Service.
      Trade, industry and taxation in Liberia.

   HL   1963   HF3939
                                  L4A3

CA   LIBERIA.
      The Liberia official gazette includes a few data from time to time, e.g., annual foreign trade statistics, lists of passports issued, etc.

CA   LIBERIA. Customs Service.
      ... Annual report ....
      Report year ends Sept. 30th.

   1922/23-1924/25   336.26
                                  L695

CA   LIBERIA. Customs Service.
      Import, export and shipping statistics ... with comparative tables. 1924-
      Statistics from 1920.

CA   **LIBERIA. Bureau of Statistics.
      Trade statistics showing imports from and exports to foreign countries ... with comparative statements for the same period 1919/22//

CA   **LIBERIA. Bureau of Statistics.
      Statistics relating to shipping, import and export trade and customs revenue of Liberia 1905/08//

CA   **LIBERIA. Customs Service.
      Preliminary statement of trade statistics ....

CA   **LIBERIA. Customs Service.
      Supplemental statement of trade statistics ....

CB   LIBERIA. Bureau of Statistics.
CA   F.T. [i.e., foreign trade] report.
      Issued in 4 parts each year, 101, 102, 201, 202.
      Each part issued quarterly through 1957; 1958- annually.
      Supplements and summary reports with some years.

   1957-63+   382.666
   wanting: 1960 and several   L695f
     other parts
   Suppl. and summary reports:
     1953, 1953-55, 1953-56,
     1953-58, 1954-59

AR   LIBERIA. Dept. of Agriculture and Commerce.
      Report on the operation of the Department of Agriculture and Commerce.

   1960/61+   S365
                                  A3

LIBYA

SA   LIBYA. Census and Statistical Dept.
      Statistical abstract.
      English and Arabic.

   1958-62 (1 vol.),   HA2167
   1963-64+   L5A3

SB	LIBYA. Census and Statistical Dept. *Statistical summary.* Quarterly. English and Arabic.		SA	LITHUANIA. Finansu Ministerija. *... Lietuva skaitlinemis. Diagramu albomas. La Lithuanie en chiffres. Album de diagrammes.* 1924.	
	no. 22/24-100/101, Oct./Dec. 1955- Apr./June 1962+ wanting: no. 94/96, Oct./Dec. 1961	HA2167 L5A35		HL 1924	HC337 L5L775

SB **LITHUANIA. Centralinis Statistikos Biuras.
 Statistikos biuletenis. Bulletin de statistique. no. 1, Nov. 1923-
 Lithuanian, French, and English.
 At head of title: Finansu Ministerija.

CA	LIBYA. Census and Statistical Dept. *External trade statistics.*				
	1955-64+ wanting: 1963	HF270 L5A3			

CA **LITHUANIA. Centralinis Statistikos Biuras.
 Lietuvos užsienio prekyba. Commerce extérieur de la Lithuanie.

CB	LIBYA. Census and Statistical Dept. *Summary of external trade statistics.* 1956-	
	1956:1-2, 1962: 2, 1963: 1-2, 1964: 1, 3, 1965: 1+	HF270 L5A32

LUXEMBURG

LIECHTENSTEIN

SA	LUXEMBURG (GRAND DUCHY). Office de la Statistique Générale. *Annuaire statistique.* 1955-	
	1955, 1960, 1964-65+	314.939 L977

SA LIECHTENSTEIN.
 Rechenschafts-Bericht der Fürstlichen Regierung an den hohen Landtag. Report of the Princely Government to the High Diet. 1922-

LITHUANIA

SA	LUXEMBURG (GRAND DUCHY). Office de la Statistique Générale. *Annuaire officiel.* 1915- Published every 3 years. Suspended 1941-45. Issued by the Office under variant names: 1915-23, Commission Permanente de Statistique; 1924-40, Office de Statistique.	
	1952, 1955, 1958, 1961+	354.4939 L977

SA	LITHUANIA. Finansu Ministerija. *Lietuvos statistikos metrastis.* Statistical yearbook of Lithuania.	
	HL 1936	

SA CA	GREAT BRITAIN. Dept. of Overseas Trade. *... Economic conditions in Lithuania ... Report ....* [1924-35.] Title varies.	
	1924, 1928, 1930, 1931, 1935	380.005 F786

SA	LUXEMBURG. Service Central de la Statistique et des Etudes Economiques. *Statistiques économiques luxembourgeoises; résumés rétrospectifs.*	
	1949	330.94939 L977

SA   GREAT BRITAIN. Dept. of Overseas
CA   Trade.
       ... Report on the economic and commercial conditions in the Grand Duchy of Luxembourg .... 1921.

      1921   380.005
                                           G786

SA   LUXEMBURG. Statistisches Amt.
       Publikationen 1, 1902-
      Part of set is issued in French, part in German. The following are the principal serial publications numbered in this series: Mouvement de la population; Note statistique; Recensement du bétail; Statistik der Landwirtschaft.

SB   LUXEMBURG. Service Central de la Statistique et des Etudes Economiques.
       Bulletin du STATEC.
      Issued Mar. 1955-July 1962 by the Service under its earlier name: Service d'Etudes et de Documentation.
       Title varies: Mar. 1955-June 1963, Bulletin économique.
       Absorbed Bulletin statistique de l'Office de la Statistique Générale, July 1963.

      v. 7:12-v. 12:2,   HA1411
        Dec. 1961-Feb. 1966+ A35

SB   LUXEMBURG. Office de la Statistique Générale.
       Bulletin statistique. v. 1-11, 1950-60; n.s. no. 1-11, July 1961-Jan. 1963//
       Title varies: 1950-54, Bulletin du Service d'Etudes et de Documentation Economiques et de l'Office de la Statistique Générale.
       Supersedes its Statistique économiques luxembourgeoises.
       Merged with Bulletin économique du Service d'Etudes et de Documentation to form Bulletin du STATEC.

      v. 1-11, n.s. no. 1-11,   330.94939
        1950-Jan. 1963   L977b

CB   LUXEMBURG.

      see BELGIUM CB

AR **LUXEMBURG.
       Rapport général sur la situation agricole ....
      Volumes are issued in the series Statistique historique du grand-duché de Luxembourg covering 1839/89, 1883/93, 1893/96, 1896/1900.

M

MACAO

SA   MACAO. Repartição Central dos Serviços de Administração Civil.
       Anuário estatístico de Macau.

      1951, 1962-64+   HA1950
                                     M3A3

SB **MACAO. Secção de Estatística.
       Boletim. 1906-

CA   MACAO. Serviços de Economia e Estatística Geral.
       Comércio externo. 1, 1955-

      1962-64+   HF259
                                   M3A3

MADAGASCAR

see also MALAGASY REPUBLIC
         FRANCE (OVERSEAS)

SA   FRANCE.
       Madagascar 1954.

      1954   HA2092
                                   A3

SA   MADAGASCAR. Service de la Statistique Générale.
       Annuaire statistique du Madagascar.

      1938-51

SA **MADAGASCAR.
    Annuaire général de Madagascar et
dépendances. 1892-
    1892-1913 as Guide annuaire ....

SB MADAGASCAR. Service de Statistique
        Générale.
    Bulletin de statistique générale de
Madagascar et dépendances. 1, 1949-
    Quarterly; no more published after
no. 24, 1954.

    13-24, 1952-54   HA2091
                                A3

SB MADAGASCAR.
    Bulletin économique; publié par les
soins du gouvernement générale 1,
1900-
    Continues Notes, reconnaissances et
explorations.
    Suspended 1916-18.
    1923-25 in two parts: Partie: Docu-
mentation générale and Partie: Statistiques. In 1923, Partie: Statistiques has
no. 1-2/3 called Supplément; 1924, only
no. 3-4; 1925, no. 1-4.
    Continued as Bulletin économique
mensuel ....

    1925 and Suppl.; 1925/26;   330.969
    1927, pt. 1; n.s. 24-26,   M178
    Nov. 1926-Jan. 1929

CA MADAGASCAR.
    Statistiques du commerce et de la
navigation. Régime douanier de Madagascar.

    1950-52   382.691
                                M178

MALAGASY REPUBLIC

see also MADAGASCAR
        FRANCE (OVERSEAS)

SA MALAGASY REPUBLIC.
    Annuaire national. 1962-

    HL 1963+   JQ3451
                                A1A3

SB MALAGASY REPUBLIC. Service de
        Statistique et des Etudes Socio-
        économiques.
    Bulletin mensuel de statistique. no. 1,
Oct. 1955-
    Issued 1955-60 by Madagascar. Ser-
vice de Statistique Générale.

    no. 18-106, Mar. 1957-   HA2091
    July 1964   A35
    wanting: several issues

CA MALAGASY REPUBLIC. Service de
        Statistique et des Etudes Socio-
        économiques.
    Statistiques du commerce extérieur de
Madagascar. 1957-

    1958-64+   HF268
    wanting: 1960   M2A4

CA MALAGASY REPUBLIC. Service de
        Statistique et des Etudes Socio-
        économiques.
    Statistiques du commerce extérieur de
Madagascar; séries rétrospectives, 1949-
1961.

    1949-61 (1 vol.)   HF268
                                M2A39

MALAWI

see also NYASALAND before 1964
        RHODESIA AND NYASALAND

SB MALAWI. Ministry of Finance.
    Quarterly digest of statistics. Apr.
1964-
    Supersedes in part the Monthly digest
of statistics of the Federation of Rhodesia
and Nyasaland, published by the Central
African Statistical Office.

    Apr. 1964-Apr. 1965+   HA1977
                                M3

CA MALAWI. Central Statistical Office.
    Annual statement of external trade.

    1964

AR MALAWI. Dept. of Agriculture.
    Annual report.

    FRI 1962-63, pt. 1-2+

MALAYSIA

Publications are listed as follows:

STRAITS SETTLEMENTS (1858-1946)
MALAY STATES
MALAY STATES, FEDERATED
MALAYA (FEDERATION), 1948-
MALAYSIA, 1963-

see also SINGAPORE
         NORTH BORNEO
         SABAH
         SARAWAK

STRAITS SETTLEMENTS

SA GREAT BRITAIN. Colonial Office.
    Colonial reports--annual. Straits
    Settlements.

    1890-1938   325.342
                                 G787

SA **STRAITS SETTLEMENTS. Statistical
    Office.
    Short statistics: summary.

SA STRAITS SETTLEMENTS.
    Blue book, for the year 1867-

    1912, 1917-38   315.95
                                 S896

SA MALAYA. Statistics Dept.
    Malayan yearbook.

    1935

SA STRAITS SETTLEMENTS.
    Annual departmental reports of the
    Straits Settlements for the year ....
    1911-

    1911, 1920-23   354.595
                                 S896

CA STRAITS SETTLEMENTS. Statistical
    Office.
    The foreign trade of Malaya.

    1935, 1937

CA **STRAITS SETTLEMENTS. Registrar of
    Imports and Exports.
    Summary of foreign imports and
    exports. 1915-20//

CA **STRAITS SETTLEMENTS. Registrar of
    Imports and Exports.
    Annual summary of monthly return of
    foreign imports and exports of British
    Malaya.

CA **STRAITS SETTLEMENTS. Registrar of
    Imports and Exports.
    Returns of imports and exports.
    Annual.

CB **STRAITS SETTLEMENTS. Registrar of
    Imports and Exports.
    British Malay. Return of foreign
    imports and exports. July 1921+
    Combined tables of Return for Straits
    Settlements, Federated Malay States and
    Non-Federated Malay States. Separate
    tables for Straits Settlements are also
    published separately in folio.

CB **STRAITS SETTLEMENTS. Registrar of
    Imports and Exports.
    Return of imports and exports ....
    Half-year.

CB **STRAITS SETTLEMENTS. Registrar of
    Imports and Exports.
    Return of imports and exports. Quar-
    terly.   -Dec. 1923//

AA STRAITS SETTLEMENTS. Dept. of Agri-
    culture.
    Malayan agricultural statistics.

    FRI 1931-40, 1947-49

MALAY STATES

SA   MALAY STATES.
     Report relating to ....

     1890-95   Brit. Docs.

SA   BRUNEI.
     Report on the state of Brunei by the British Resident.

     HL   1914-15, 1917-23

     see also BRUNEI

SA   KEDAH.
     Annual report of the Advisor to the Kedah government and the annual report of the Advisor to the Perlis government.

     1910-18   Brit. Docs.

SA   KELANTAN.
     Administration report by the British Advisor.

     1909-18   Brit. Docs.

SA   NEGRI SEMBLIAN.
     Administration report by the British Resident.

     HL   1914-23

SA   PAHANG.
     Administration report by the British Resident.

     HL   1914-23

SA   PERAK.
     Administration reports by the British Resident.

     HL   1914-23

SA   SELANGOR.
     Administration reports by the British Resident.

     HL   1914-23

SA   TRENGGANU.
     Annual report of the British Agent.

     1910-18   Brit. Docs.

     HL   1914, 1917-22

MALAY STATES, FEDERATED

SA   GREAT BRITAIN. Dept. of Overseas
CA   Trade.
     ... Report on economic and commercial conditions in Malaya .... [1928-39.]
     Title varies: 1928-31, A review of the trade of British Malaya ....

     1928, 1931, 1934,   380.005
       1937, 1939   G786

SA   MALAY STATES, FEDERATED.
     Manual of statistics relating to the Federated Malay States.

     1908, 1915, 1920-21,   315.95
       1923, 1926   M239m

SA   MALAY STATES, FEDERATED.
     [Annual] report ....
     For later reports see Gt. Brit. Colonial Office. Colonial reports--annual (325.342 G787).

     1914-21   325.342
                                G786

CA   MALAY STATES, FEDERATED. Customs and Excise Dept.
     Report on the Customs and Excise Department for the year ....

     1939

CA   MALAY STATES, FEDERATED. Trade and Customs Dept.
     ... Report on the Trade and Customs Department.
     Supplement to the F.M.S. government gazette.

     1914-23   382.91
                                M238

CB **MALAY STATES, FEDERATED. Trade
        and Customs Dept.
        Quarterly import and export returns.
    1918-

AA MALAY STATES, FEDERATED. Dept.
        of Agriculture.
        Malayan agriculture. Handbook com-
    piled by the Department of Agriculture,
    Federated Malay States and Straits Settle-
    ments.

    1922, 1924   630.6595
                                  M239m

AB MALAY STATES, FEDERATED. Dept.
        of Agriculture.
        The Malayan agricultural journal.

        FRI   v. 11: no. 10-v. 29:
                no. 11, Oct. 1923-
                Nov. 1941
                Index, 1912-35

MALAYA (FEDERATION)

SA MALAYA (FEDERATION).
        Official year book. 1, 1961-
        Includes detailed statistical appendix.

    1962+   959.5
                                  M238

SA MALAYA (FEDERATION).
        Report. 1946-56.
        Began publication with the report for
    Apr./Dec. 1946.
        Reports for Apr./Dec. 1946-47 issued
    by the Malayan Union.
        Issued also in the series Colonial
    annual reports (Gt. Brit. Colonial Office).
        Preceded by Annual report on the
    social and economic progress of the
    people of the Federated Malay States
    issued by Gt. Brit., Colonial Office, in
    its numbered series, Colonial reports--
    annual (325.342   G787).

    1946-56   325.342
                                  G7871Ma

CA MALAYA (FEDERATION). Dept. of Sta-
        tistics.
        Statistics of external trade. Annual.
        Issued 1921-30 by British Malaya;
    1947-   by the Malayan Union.
        Title varies: 1921-30, Return of for-
    eign imports and exports; 1947-58,
    Imports and exports (including trade with
    Singapore and foreign countries).

    1926, 1929-30, 1947-50,   382.595
      1952, 1962-63+   M239

CB MALAYA (FEDERATION). Dept. of Sta-
        tistics.
        Monthly statistics of external trade.
    1961-

AR MALAYA (FEDERATION). Dept. of
        Agriculture.
        Annual report.

        FRI   1914-15, 1918-23,
                1937-38, 1947-62,
                1964

MALAYSIA

SA MALAYSIA. Dept. of Statistics.
        Bulletin of statistics, Malaysia.
    Annual.

SB MALAYSIA. Dept. of Statistics.
        Statistical bulletin of the states of
    Malaya.
        Published through Sept. 1963 by the
    Dept. of Statistics of the Federation of
    Malaya, with title: Statistical bulletin of
    the Federation of Malaya.

    Jan. 1957-Feb. 1966+   315.95
    wanting: several issues   M239s

CB MALAYSIA. Dept. of Statistics.
        Malaysian quarterly statistics of exter-
    nal trade.

AB MALAYSIA. Ministry of Agriculture and Cooperatives.
    The Malaysian agricultural journal.
    Issues -v. 44: no. 1, 1963, have title The Malayan agricultural journal, issued by the Federation of Malaya.

    FRI 1947-65+

    see also MALAY STATES, FEDERATED (AB)

MALI

see also SUDAN, FRENCH before 1960

SA Bamako. Chambre de Commerce, d'Agriculture et d'Industrie.
    Annuaire statistique de la République du Mali.

    1962+   HA2117
                         M3B3

SA MALI. Direction de la Statistique.
    Annuaire statistique.

SB MALI. Service de la Statistique Générale de la Comptabilité Economique Nationale.
    Bulletin mensuel de statistique. 1960-
    Issued 1960-63 by the Service under an earlier name: Service Statistique.
    Supersedes Bulletin statistique mensuel formerly issued by Sudan, French. Service Statistique.

    Mar. 1960-Jan./Feb. HA2117
    1966+   M3A3
    wanting: several issues

SB MALI. Service Fédéral de la Statistique et de la Mécanographie.
    Bulletin statistique et économique mensuel. 1, 1959-
    Supersedes bulletin with the same title published by French West Africa. Service de la Statistique Générale.

MALTA

SA MALTA. Office of Statistics.
    Statistical abstract of the Maltese Islands. 1946-

    1946-64+   HA1117
    wanting: 1962   M3A3

SA MALTA. Office of Statistics.
    Malta statistical handbook. 1, 1965-

    1965+   HA1117
                         M3A32

SA GREAT BRITAIN. Dept. of Overseas
CA Trade.
    ... Report on economic conditions in Cyprus and Malta. 1935.

    1935   380.005
                         G786

SA GREAT BRITAIN. Colonial Office.
    Malta, annual report. 1890/91-1920/21.
    Issued in its numbered series, Colonial reports--annual.

    1890/91-1919/20   Brit. Docs.
    1920/21 (no. 1111)   325.342
                         G787

SA MALTA.
    Blue book. 1821(?)

SB MALTA. Office of Statistics.
    Quarterly digest of statistics. 1, Mar. 1960-

    Mar. 1960-Dec. 1965+ HA1117
                         M3A35

SB MALTA. Office of Statistics.
    Statistical summary for the month.

    Mar., Apr., Nov. 1965- HA1117
    May 1966+   M3A36

CA   MALTA. Office of Statistics.
       The trade of the Maltese Islands.
     Annual. 1876-
       Compiled in collaboration with the
     Customs and Port Dept.

     1951-64+   382.458
     wanting: 1962   M261t

CA   MALTA.
       Trade statistics. Customs. (?)-
     1950//
       1951- continued in the Trade of the
     Maltese Islands.

     1874, 1904/05-1907/08,   382.458
        1946-50   M261

CB   MALTA. Office of Statistics.
       The trade of the Maltese Islands.
     Quarterly.
       Compiled in collaboration with the
     Customs and Port Dept.

     1951-64+   382.458
                                  M261ta

AA   MALTA. Office of Statistics.
       Census of agriculture.
       Title varies: 19 -60, Report of the
     census of agriculture.
       None published 1962.

     1957, 1959-64+   S269
                                  M3A4

MANCHURIA

CA   MANCHURIA. Dept. of Finance.
       Annual returns of the foreign trade of
     Manchoukuo.

     1932-34, 1935: pt. 1,
        1936: pt. 1, 1937:
        pt. 1-2, 1938: pt. 1-2

CB   MANCHURIA. Dept. of Finance.
       Monthly returns of the foreign trade of
     Manchoukuo.

     Jan. 1933-Dec. 1938   382.518
                                  M268

MARTINIQUE

SA   FRANCE. Institut National de la Statis-
     tique et des Etudes Economiques.
       Annuaire statistique de la Martinique.

     1952-56, 1956-59,   317.298
        1959-60   F815

SA   **MARTINIQUE.
       Annuaire de la Martinique. 1860-
     1917//
       Continues Almanach de la Martinique.

SA   **MARTINIQUE.
       Almanach de la Martinique. 1825-59//
       Continued as Annuaire de la Marti-
     nique.

AB   **MARTINIQUE. Service de l'Agriculture.
       Bulletin agricole.
       1898-99 published under patronage of
     Comité Consultatif du Jardin Botanique.

MAURITANIA

see also FRENCH WEST AFRICA

SA   MAURITANIA. Service de la Statistique.
SB   Bulletin statistique et économique. 1,
     1960-
       Title varies slightly.

     1960: no. 1-1964: no. 7+   HA2117
                                  M4A3

CB   MAURITANIA. Service de la Statistique.
       Commerce extérieur: produits/pays,
     pays/produits. 1966(?)- Quarterly.

MAURITIUS

SA   GREAT BRITAIN. Colonial Office.
       Annual report on Mauritius. 1946-
       Previously issued in its numbered
     series, Colonial reports--annual

SA   MAURITIUS. Central Statistical Office.
        Year book of statistics. no. 1-14, 1946-59//
        Continued by its Digest of statistics (SB).

1947-59	316.982 M456

FRI 1949-52

(325.342 G787) no. 1115: 1920-
no. 1905: 1938. Suspended 1940-46.

1946-63+	325.342 G7871M

SA   MAURITIUS.
        Blue book.
        Discontinued 1947.

1914-24, 1928, 1945-47	354.6982 M456

SB   MAURITIUS. Central Statistical Office.
        Digest of statistics. 1, Mar. 1961-
        Quarterly.
        Continues its Year book of statistics (SA).

Mar. 1961-Sept. 1965+	316.982 M456d

CA   MAURITIUS. Customs and Excise Dept.
        Report.
        Reports for 1946-52 issued by the department under its earlier name: Customs Dept.
        Title varies: -1947, Report on trade.

1946-64+	336.26 M456

CB   MAURITIUS. Customs Dept.
        Statistical report. Quarterly.

AR   MAURITIUS. Dept. of Agriculture.
        Report.

1947, 1954-63+	S338 M4A3

AB **MAURITIUS. Dept. of Agriculture.
        Statistical series. Bulletin. English edition. 1-6, 1914-21//

AB **MAURITIUS. Chamber of Agriculture.
        Bureau of agricultural statistics and information... Bulletin. no. 1-46, 1909/13//

MEXICO

SA   MEXICO. Dirección General de Estadística.
        Anuario estadístico de los Estados Unidos Mexicanos. 1893-

1893-95, 1899, 1901-03, 1920-26, 1930, 1938, 1940-45, 1953-57, 1960-61--1962-63+	HA761 A3

SA   MEXICO. Dirección General de Estadística.
        Compendio estadístico.

1941, 1947, 1951-57	HA761 A35

FRI 1960

SA   GREAT BRITAIN. Dept. of Overseas
CA   Trade.
        ... Economic conditions in Mexico ... Report .... [1920-36.]
        Title varies.

1920-21, 1922-23, 1927, 1931, 1933, 1936	380.005 G786

SA   MEXICO. Dirección General de Estadística.
        Cuador sinóptico y estadístico de la República Mexicana formado por la Dirección ....

1900	317.2 M612c

SA **MEXICO. Dirección General de Estadística.
        Estadística general de la República Mexicana. 1-10, 1884-96//

SA   MEXICO. Ministerio de Fomento, Colonización e Industria.
        Memoria.

      1856/57, 1867/68-1868/69,   330.972
        1872/73, 1883/85: 2,   M61
        1892-96

SB   MEXICO. Dirección General de Estadística.
        Revista de estadística. Mar. 1938-
        Supersedes Revista de economía y estadística.

      Mar. 1938-Sept. 1965+   317.2
      wanting: several issues   M610

SB   MEXICO. Secretaría de la Economía Nacional.
        Revista de economía y estadística ....
    May 1933-

      v. 1-4, 1933-36   317.2
                              M64

SB   MEXICO. Departamento de la Estadística Nacional.
        Estadística nacional .... no. 1-121/22;
    Jan. 15, 1925-Nov./Dec. 1932//

      1-121/22   317.2
      wanting: 62-65, 70   M63

SB **MEXICO. Departamento de la Estadística Nacional.
        Boletín. Ser. 2, año 1-2, July 1923-
    Dec. 1924//
        Continued as Estadística nacional.
        Continues Dirección General de Estadística. Boletín.

SB **MEXICO. Dirección General de Estadística.
        Boletín. 1-5, 1912-14//
        Continued under Departamento de la Estadística Nacional.

CA   MEXICO. Dirección General de Estadística.
        Anuario estadístico del comercio exterior de los Estados Unidos Mexicanos. 1920-
        1939 report covers 6-year period, 1934-39.
        Title varies.

      1940, 1942-50, 1952,   382.72
        1962-63+   M6a

CA   MEXICO. Dirección General de Estadística.
        Anuario estadístico del comercio de los Estados Unidos Mexicanos con los paises de la asociación latinoamericana de libre comercio.

CA **MEXICO. Departamento de la Estadística Nacional.
        Resumen del comercio exterior y navegación. no. 1-8, May-Dec. 1920;
    n.s. 1, 1925.
        Continues Ministerio de Hacienda y Crédito Público. Resumen de la importación.
        No. 1-8 issued by Departamento de Legislación y Estadística ....

CA   MEXICO. Secretaría de Hacienda y Crédito Público.
        ... Comercio exterior y navegación
    .... 1896/97-1899/1900, Jan.-June 1921.
        Continuation of its Estadística fiscal. Exportación and Estadística fiscal. Importación.
        Publication suspended 1900-20 inclusive.
        Superseded by Mexico. Dirección General de Estadística. Anuario estadístico comercio exterior y navegación.

      1898-99   HF131
                                 A5

CA   MEXICO. Secretaría de Hacienda y Crédito Público.
        Anuario de estadística fiscal. 1911-12--
        Continuation of its Boletín de estadística fiscal.
        Continued in the Anuario estadístico comercio exterior y navegación.

      1912-13, 1918-19   382.72
                                 M611a

CA **MEXICO. Dirección General de Estadística.
    <u>Importación y exportación</u>. 1889-1907//

CA **MEXICO. Ministerio de Hacienda y Crédito Público.
    <u>Importación</u>. 1884/85-1895/96//
    To 1885/86 as its <u>Noticias de las mercancías importadas</u>. Continued as its <u>Comercio exterior y navegación</u>.

CA **MEXICO. Ministerio de Hacienda y Crédito Público.
    <u>Exportación</u>. 1872/75-1895-96//
    1872/75-1888/89 as its <u>Noticia de la exportación de mercancías</u> (title varies slightly); 1889/90-1892/93, <u>Exportaciones</u>. Continued as its <u>Comercio exterior y navegación</u>.

CB   MEXICO. Secretaría de Hacienda y Crédito Público.
    <u>Boletín de estadística fiscal</u>.
    Continued by <u>Anuario de estadística fiscal</u>.

no. 273-284, 289-294	382.72
wanting: 275, 292	M611bo

CB   MEXICO. Dirección General de Estadística.
    <u>Resumen del intercambio comercial con los paises de la asociación latino-americana de libre comercio</u>.

CB **MEXICO. Ministerio de Industria, Comercio y Trabajo.
    <u>Boletín de industria, comercio y trabajo</u>. v. 1-5, no. 6, July 1918-Dec. 1920//
    v. 1, no. 1, in 3 editions: Departamento de Comercio, <u>Boletín de comercio</u>; Departamento de Industrias, <u>Boletín de industrias</u>; and Departamento de Trabajo, <u>Boletín del trabajo</u>.
    Continued as Departamento de Industrias, <u>Boletín de industrias</u>; Departamento de Comercio, <u>Boletín comercial</u>; and Departamento del Trabajo, <u>Boletín mensual</u>.

AA   MEXICO. Departamento Agrario.
    ... <u>Memoria del Departamento Agrario ... Apéndice estadística</u> ...

    FRI   1935/36, Sept. 1957-Aug. 1958

AA   MEXICO. Dirección General de Agricultura.
    <u>Producción agrícola</u>.

1925-27 (1 vol.)	630.672
	M61p

AA   MEXICO. Comisión Nacional Agraria.
    <u>Estadística</u>.

1915-27	630.972
	M612

AB   MEXICO. Dirección General de Economía Rural.
AA
    <u>Boletín mensual de estadística agrícola</u>. 1928-

    FRI   no. 2, 1928-65--
        wanting: several issues

AB   MEXICO. Dirección General de Agricultura.
    <u>Resumen agrícola</u>. 1, Apr. 1932-

1-10, Apr. 1932-Jan. 1933	630.672
	M61r

AB   MEXICO. Dirección General de Agricultura.
    <u>Boletín</u>. 1911-Feb. 1915//
    1911 published in 3 parts as Revista de agricultura, Revista de economía rural y sociología, and Revista forestal; 1912 published in 2 parts as Revista agricultura and Revista de economía rural y sociología.
    Continued by <u>La revista agrícola</u>.

1911-12	
wanting: no. 4, 8-9	630.672
	M61bo

MONTSERRAT

SA   GREAT BRITAIN. Colonial Office.
         Montserrat; report. 1955-56--
      Continues in part its Report for the
      Leeward Islands, issued in the numbered
      series of Colonial reports (325.342
      G7871Le).

      1957-62+   325.342
                                        G7871Mo

SA **MONTSERRAT.
         Blue book. 1826-87//
         1821-25 in Antigua. Blue book.
         Continued in Leeward Islands. Blue
      book.

AA   WEST INDIES. Federal Statistical Office.
         Agricultural statistics, series 1.
      (1956-58 survey)

      FRI   no. 2. The survey in
               Montserrat. 1959

MOROCCO

SA   MOROCCO. Service Central des Statis-
         tiques.
         Annuaire statistique du Maroc. 1925-
      Vols. for 1925- issued by the Direc-
      tion de l'Agriculture, du Commerce et
      des Forêts under an earlier name: Direc-
      tion Générale de l'Agriculture, du Com-
      merce et de la Colonisation; 19 -38, by
      the Service du Commerce et de l'Indus-
      trie; 1939-44, by the Service under an
      earlier name: Service des Statistiques.
         Title varies: 1925-49, Annuaire sta-
      tistique de la zone française du Maroc.

      1938, 1939/44, 1945/46,   HA2181
      1947/48, 1949, 1955/56,   A2
      1958-61+

SA   MOROCCO. Service Central des Statis-
         tiques.
         Tableaux économiques du Maroc,
      1915-1959.

      1915-59 (1 vol.)   330.964
                                        M868

SA   GREAT BRITAIN. Dept. of Overseas
CA   Trade.
         ... Survey of economic conditions in
      Morocco .... [1919-39.]
         1919/20 issued in the series of parlia-
      mentary papers as Parliament. Papers
      by command. Cmd. 975.
         Statistics for 1922 are found in the Sur-
      vey of the economic and commercial con-
      ditions in Algeria, Tunis, and Morocco,
      1921-22.
         Title varies.

      1920, 1921, 1923-33,   380.005
         1935, 1936, 1939   G786

SA   MOROCCO.
         Annuaire économique et financier.
         At head of title: Gouvernement Cheri-
      fien. Protectorate Français au Maroc.

      HL   1917-1918/19,   HC591
              1921/22   M8A2

SB   MOROCCO. Service Central des Statis-
         tiques.
         Bulletin mensuel de statistique. Sept.
      1957-

      no. 38, Jan. 1961-no. 90,   316.4
         Oct. 1965+   M867b

SB   MOROCCO. Service Central des Statis-
SA   tiques.
         La conjoncture économique marocaine.
      Jan. 1947-Dec. 1956//
         Monthly, with separate annual summary.
      Superseded by the service's Bulletin
      mensuel de statistique in Sept. 1957.

      Mar. 1952-Jan. 1954;   HC591
         Annual summary, 1952   M8A35

CA   MOROCCO. Direction du Commerce et de
         la Marine Marchande.
         Statistiques du mouvement commercial
      et maritime du Maroc. 1905-
         Issued by Comité des Douanes, Tan-
      gier, 1905-22; by Service du Commerce
      et de l'Industrie, 1923-38.
         Title and agency vary slightly.

      1933, 1939, 1940/45,   382.64
         1946-63--   M867

      HL   1905-39
              wanting: several issues

CA	MOROCCO. Direction du Commerce et de la Marine Marchande. <u>Le commerce extérieur du Maroc.</u> 1912-54 (1 vol.)	382.64 M867c	SB	MOZAMBIQUE. Repartição Técnica de Estatística. <u>Boletim trimestral de estatística.</u> 1925-47// 1925-39 monthly under title <u>Boletim econômico e estatístico.</u>	
CB	MOROCCO. Direction du Commerce et de la Marine Marchande. <u>Note de documentation.</u> no. 79-326, Sept. 15, 1951-Feb. 1966+ wanting: no. 184, 1958	382.64 M867n	CA	MOZAMBIQUE. Direcção dos Serviços de Estatística Geral. <u>Comércio externo.</u> <u>Commerce extérieur.</u> Agency name varies. 1958- issued in two volumes. 1957-62, v. 1+	382.679 M939

MOZAMBIQUE

SA	MOZAMBIQUE. Direcção dos Serviços de Estatística Geral. <u>Anuário estatístico.</u> <u>Annuaire statistique.</u> Began publication in 1929 with volume for 1926/28. Supplements accompany some issues. 1937, 1949, 1950-63+	HA2009 M7A35	CA	MOZAMBIQUE. Circulo Aduaneiro. <u>Estatística do comércio e navegação.</u> 1885/87- 1909-12, 1919-20, 1933, 1935, 1938	382.679 M938
			AA	MOZAMBIQUE. Direcção dos Serviços de Estatística Geral. <u>Estatística agrícola.</u> 1941- FRI 1942, 1944, 1949-61+	
SA CA	GREAT BRITAIN. Dept. of Overseas Trade. <u>... Report on economic and commercial conditions in Portuguese East Africa ....</u> [1921-38.] Title varies. 1921, 1927, 1929, 1932, 1935, 1938	380.005 G786	AB	MOZAMBIQUE. Repartição de Agricultura. <u>Bulletin.</u> Ser. 1-2: 1-21, 1909-14; Ser. 3: no. 1-12, Jan.-Dec. 1915// no. 1-4, 1909-10	630.6679 M939
SA	**PORTUGAL. Ministério das Colónias. <u>Relatórios e informações.</u> 1, 1917// Contains material on Angola.		AR	MOZAMBIQUE. Direcção dos Serviços de Agricultura. <u>Relatório.</u> 1940-44-- Supersedes Mozambique. Repartição de Agricultura. <u>Relatório.</u> 1940-44 (1 vol., 4 parts)	630.6679 M9391
SB	MOZAMBIQUE. Direcção dos Serviços de Estatística Geral. <u>Boletim mensal.</u> <u>Bulletin mensuel de statistique.</u> <u>Monthly digest of statistics.</u> ano 1, 1960- v. 4: no. 1-v. 6: no. 11, 1963-Nov. 1965+	HA2209 M7A4			

N

NATAL

SA **NATAL. Native Affairs Dept.
    Blue book on native affairs.
    Compiled from reports by magistrates and from records in the office of the Secretary for Native Affairs. Continued as Union of South Africa. Native affairs Dept.

CA NATAL. Customs and Excise Dept.
    Report of the collector of customs.
    Title varies.

    1908   336.26
                                          N271

AA **NATAL. Dept. of Agriculture.
    Agricultural statistics of Natal.
    1905/06//

AB NATAL. Dept. of Agriculture.
    Bulletin. no. 1-18, 1902/09//

    no. 4, 12 (1904-06)   630.6684
                                          N271

NAURU

SA AUSTRALIA. Dept. of Territories.
    Report to the General Assembly of the United Nations on the administration of the Territory of Nauru.
    Supersedes the report to the Council of the League of Nations on the administration of the Territory of Nauru, issued 1932-35, 1937-40, by the Nauru Administrator.

    1947/48-1958/59 (later   354.997
      issues are included in the   N312a
      bound set of Australian
      Parliamentary Papers)

    HL   1947-58   DN553
                                          N3A93

SA NAURU. Administrator.
    Report on the administration of Nauru during the year ....
    Prepared by the administrator for submission to the League of Nations. 1915/20-
    Superseded by the Report to the General Assembly of the United Nations on the administration of the Territory of Nauru, issued by Australia. Dept. of Territories.

    HL   1915/20-1940   DU553
                                          N3

NETHERLANDS

SA NETHERLANDS. Centraal Bureau voor de Statistiek.
    Statistisch zakboek. 1926- (None issued 1927.)
    An earlier publication with same title covers the period 1899-1924.

    1944-59, 1962-65+   314.92
    wanting: 1949   N468sz

SA NETHERLANDS. Centraal Bureau voor de Statistiek.
    Jaarcijfers voor Nederland. 1881-
    Title and agency vary.
    Vols. for 1881-1921 published in two parts: [1] "Binnenland" (after 1897, "Rijk in Europa"); [2] "Koloniën." The part designated as "Koloniën" (314.92 N468K) was discontinued after 1921, the statistics being continued in a publication issued by the Dutch East Indies Departement van Landbouw, Nijverheid en Handel with title: Statistisch jaar overzicht voor Nederlandsch-Indië ... (Statistical abstract for the Netherlands East Indies).

    1892-93, 1896, 1898-   314.92
      1941/42, 1943/46,   N468
      1947--1961-62+

SA NETHERLANDS. Centraal Bureau voor de Statistiek.
    Statistiek in tijdreeksen: zestig jaren statistiek in tijdreeksen, 1899-1959.

    FRI   1899-1959

SA NETHERLANDS (KINGDOM, 1815-   ).
    Departement van Overzeesche
    Gebiedsdeelen.
    Indisch verslag. I. Tekst van het
    verslag van bestuur en staat van
    Nederlandsch-Indië over het jaar. 1930-
    Continues in part Verslag van bestuur
    en staat van Nederlandsch-Indië, Suri-
    name en Curaçao, issued by the same
    office.
    Statistisch jaaroverzicht van
    Nederlandsch-Indië, published by Dutch
    East Indies. Centraal Kantoor voor de
    Statistiek, since 1922/23; issued as
    Part II of the Indisch verslag, 1930-

    1930-37   325.3492
                                    N469i

SA GREAT BRITAIN. Dept. of Overseas
CA   Trade.
    ... Report on economic conditions in
    the Netherlands .... [1919-38.]
    Holland: 1919 issued in the series of
    Parliamentary papers as Papers by com-
    mand. Cmd. 872.
    Title varies.

    1919, 1921-34, 1936, 1938   380.005
                                    G786

SA NETHERLANDS. Departement van Over-
        zeesche Gebiedsdeelen.
    Verslag van bestuur en staat van
    Nederlandsch-Indië, Suriname en Cura-
    çao. 1848-1930//
    Title varies: 1848, Mededeelingen
    betreffende de koloniën; 1849-66, Ver-
    slag van het beheer en de staat der kolo-
    niën (varies slightly); 1867-1923, Kolo-
    niaal verslag.
    Continued by its Curaçaosch verslag,
    Indisch verslag, and Surinaamsch ver-
    slag.

    1911-30   JV33
                                    N2A4

SA NETHERLANDS. Centraal Bureau voor
        de Statistiek.
    Bijdragen tot de statistiek van Neder-
    land. Nieuwe volgreeks. 1900-
    Part of series catalogued separately:
    Jaarstatistiek van den in-, uit- en door-
    voer, 1917-21, no. 264, 291, 303, 323,
    343, 366, 379, 384 (382.492 N469j);

    no. 160, 170, 192, 316 catalogued as
    monographs.

    no. 1, 117, 145, 187, 191,   314.92
        198, 221, 222, 224,   N468sb
        232, 237-239, 250, 252,
        260, 280, 283, 286-288,
        311, 313-315, 318, 321-
        322, 326-327, 330, 339,
        341, 344

SA NETHERLANDS. Centraal Bureau voor
        de Statistiek.
    Jaarcijfers voor het koningrijk der
    Nederlanden koloniën. Statistical annual
    for the Netherlands. Part Colonies.
    1930- issued as Part II of the Indisch
    verslag (Netherlands Indian report).

    1891, 1895, 1900, 1905,   314.92
        1908-19, 1921   N468K

SA **NETHERLANDS. Centraal Bureau voor
        de Statistiek.
    Bijdragen tot de statistiek van Neder-
    land. 1-7, 1894-98//

SA **NETHERLANDS. Departement van Bin-
        nenlandsche Zaken.
    Statistisch jaarboek voor het koningrijk
    der Nederlanden 1-15, 1851-68//

SB NETHERLANDS. Centraal Bureau voor
        de Statistiek.
    Maandschrift van het Centraal Bureau
    voor de Statistiek. Revue mensuelle.
    Sept. 1906-
    Continuation of Tijdschrift van het
    Centraal Bureau voor de Statistiek.

    1909-39, 1944-64+   314.92
    wanting: some issues for   N468ms
        1915-18, 1939

    FRI Current two years
            only.

SB NETHERLANDS. Centraal Bureau voor
        de Statistiek.
    Statistisch bulletin. n.s. 1, 3n. 1945.
    Biweekly.

    FRI 1949-66+

SB NETHERLANDS. Centraal Bureau voor
    de Statistiek.
    Maandcijfers en andere periodieke
    opgaven betreffende Nederland en de
    koloniën. Nieuwe volgreeks. no. 1,
    1898-

    no. 1-35, 1898-1919   314.92
    wanting: no. 7-8   N468ma

SB **NETHERLANDS. Centraal Bureau voor
    de Statistiek.
    Tijdschrift van het Centraal Bureau de
    Statistiek .... Revue du Bureau Central
    de Statistique 1-16, 1902-06//
    Continued as Maandschrift ....

SB NETHERLANDS. Centraal Bureau voor
    de Statistiek.
    Maandcijfers en andere periodieke
    opgaven betreffende Nederland en Neder-
    landsch Oost-Indië.
    No. 1-11, eerste maanden van 1893-
    eerste maanden van 1898.

    no. 1-11, 1893-98   314.92
                                 N468m

CA NETHERLANDS. Centraal Bureau voor
    de Statistiek.
    Jaarstatistiek van den in-, uit- en
    doorvoer. Statistique du commerce des
    pays-bas avec les pays étrangers ....
    On cover, 1917-21, Bijdragen tot de
    statistiek van Nederland. Nieuwe vol-
    greeks. no. 264, 291, 303, 323, 343,
    366, 379, 383, 384.
    Title varies.
    1919- issued in 2 vols. each year.
    Continuation of Statistiek van den in-,
    uit- en doorvoer ..., issued by its
    Departement van Financiën.

    1915-26, pt. 1; 1938,   382.492
    pt. 1-2; 1939   N469j

CA **NETHERLANDS. Departement van
    Financiën.
    Statistiek van den in-, uit- en door-
    voer 1846-1916//
    Issued each year in two parts.
    1846-76 as Statistiek van den handel
    en de scheep-vaart.
    Continued as Centraal Bureau voor de
    Statistiek. Jaarstatistiek van den in-,
    uit- en doorvoer.

CA **NETHERLANDS. Departement van Land-
    bouw, Nijverheid en Handel.
    Aperçu du commerce et l'industrie des
    pays-bas .... no. 1-20, 1910//
    Published also in English.

CB NETHERLANDS. Centraal Bureau voor
CA   de Statistiek.
    Maandstatistiek van de in-, uit- en
    doorvoer per land. 1950-
    Supersedes in part its Maandstatistiek
    van de in-, uit- en doorvoer van Neder-
    land.

    1950-65+   382.492
                                 N469b

CB NETHERLANDS. Centraal Bureau voor
CA   de Statistiek.
    Maandstatistiek per goederensoort.
    1950-
    Supersedes in part its Maandstatistiek
    van de in-, uit- en doorvoer van Neder-
    land.

    1950-65+   382.492
                                 N469a

CB NETHERLANDS. Centraal Bureau voor
    de Statistiek.
    Maandstatistiek van de in-, uit- en
    doorvoer van Nederland. Jan./Mar.
    1917-Dec. 1949//
    Publication suspended Apr. 1940-July
    1945.
    Earlier statistics in Statistiek van het
    koningrijk der Nederlanden. Staten van
    de in-, uit- en doorgevoerde voornaamste
    handelsartikelen (title varies), issued by
    the Departement van Financiën, 1846-
    1918.
    Separately paged sections and supple-
    ments accompany some numbers.
    Superseded by its Maandstatistiek per
    goederensoort and its Maandstatistiek van
    de in-, uit- en doorvoer per land.

    Jan./Mar. 1917-Dec. 1949   382.492
    wanting: 1917, July 1924,   N469
    Nov./Dec. 1945

AA   NETHERLANDS. Landbouw-Economisch Institut in sammenwerking met Centraal Bureau voor de Statistiek. Landbouwcijfers.

   FRI   1954-55, 1959-60 (with English appendix), 1962-64+

AA   NETHERLANDS. Ministerie van Landbouw. Agricultural yearbook.

   FRI   1939-41

AB   NETHERLANDS. Centraal Bureau voor de Statistiek. Maandstatistiek van de landbouw. 1, Jan. 1953-

AR   NETHERLANDS. Ministerie van Landbouw, Visserij en Voldselvoorziening. Verslag. Agricultural report.

   FRI   1941, 1942-45, 1947-63+

AR   NETHERLANDS. Directie van den Landbouw.
AA   Verslagen en mededeelingen.
   Annual report and annual statistics included in this series along with technical agricultural reports.

   1908: 1, 3, 5-7   1917: 3   630.6492
   1909: 3, 5-6   1918: 5   N469v
   1910: 1, 5-6   1919: 1, 3-4
   1911: 1, 6   1920: 1, 3-4
   1912: 1, 3-5   1921: 1, 3-5
   1913: 1-6   1922: 1-4
   1914: 3-5   1943: 1-2
   1915: 1, 3-5   1945: 1
   1916: 2-5

NETHERLANDS ANTILLES

SA   NETHERLANDS ANTILLES. Bureau voor de Statistiek. Statistisch jaarboek, Nederlandse Antillen. 1956-
   Dutch, English, and Spanish.

   1963, 1965+   HA920
                                  N4A3

SA   NETHERLANDS. Departement voor Uniezaken en Overzeese Rijksdeelen. Verslag Nederlandse Antillen.
   In two parts each year: I. Tekst van het verslag van bestuur en staat van de Nederlandse Antillen. II. Statistische jaaroversicht.

   1949: 1-2, 1950: 1-2,   325.3492
   1951: 1-2, 1952: 1-2   N469w

SA   NETHERLANDS. Departement van Overzeesche Gebiedsdeelen. Curaçäosch verslag. 1932-(?)
   Issued annually in two parts: I. Tekst van het verslag van bestuur en staat van Curaçao. II. Statistische jaaroverzicht van Curaçao. 1949 entitled Verslag Nederlandse Antillen.
   Part II also has English text.

   1936-37, 1946-47   325.3492
                                  N469c

SB   NETHERLANDS ANTILLES. Bureau voor de Statistiek. Statistische mededeelingen Nederlandse Antillen. Statistical information.

   v. 12, no. 7-v. 13,   HA920
   no. 13; Jan. 1965-   N4A35
   Jan. 1966+

CA   NETHERLANDS ANTILLES. Dienst Economische Zaken en Welvaartszorg. Afdeling Statistiek. Jaarstatistiek van de in- en uitvoer per land van de Nederlandse Antillen. -1961//
   Continued in its Maandstatistiek and Kwartaalstatistiek.

   1953-61   382.72971
                                  N469j

CA   NETHERLANDS ANTILLES. Dienst Economische Zaken en Welvaartszorg. Afdeling Statistiek. Jaarstatistiek van de in- en uitvoer per goederensoort van de Nederlandse Antillen.

   1953-60   382.72971
                                  N469

CA   NETHERLANDS ANTILLES. Statistieken
CB   Planbureau.
         Maandstatistiek van de in- en uitvoer
     per goederensoort van Curaçao en Aruba.
         Annual statistics in Dec. issue.

CA   NETHERLANDS ANTILLES. Bureau
         voor de Statistiek.
         In- en uitvoer statistiek per goederen-
     soort Bonaire.

CA   NETHERLANDS ANTILLES. Bureau
CB   voor de Statistiek.
         Kwartaalstatistiek van de in- en uit-
     voer per land.
         Dec. issue contains annual figures.

NETHERLANDS EAST INDIES

see DUTCH EAST INDIES
    INDONESIA

NEVIS

see also ST. CHRISTOPHER AND NEVIS
        ST. KITTS-NEVIS-ANGUILLA

SA **NEVIS.
        Blue book. 1821-82//

AA   WEST INDIES. Federal Statistical Office.
         Agricultural statistics, series 1.
     (1956-58 survey)

     FRI   no. 3. The survey in
               Nevis. 1960

NEW CALEDONIA

SA **NEW CALEDONIA.
        Annuaire de la Nouvelle-Calédonie et
    dépendances. 1871-
        1916-21 never published.

NEW GUINEA

SA   AUSTRALIA. Dept. of External Territo-
         ries.
         Report to the General Assembly of the
     United Nations on the administration of
     the Territory of New Guinea. 1946/47-
         Report to the Council of the League of
     Nations 1914/21-

     HL   1947/48-1953,   DU742
          1956-64+   A939

SA   AUSTRALIA. Prime Minister's Dept.
         Report to the Council of the League of
     Nations on the administration of the Ter-
     ritory of New Guinea. 1914/21-
         Superseded in 1946/47 by Report to the
     General Assembly of the United Nations
     on the administration of the Territory of
     New Guinea, issued by Dept. of External
     Territories.

     HL   1914/21-1938   DU742
                                   A938

SA   AUSTRALIA. Governor-General.
         ... Report to the League of Nations on
     the administration of the Territory of New
     Guinea from July 1st, 1923, to June 30th,
     1924.

     HL   1923/24   DU742
                                   A938a
                                   1924

SA **NEW GUINEA.
         Statistics relating to commerce, native
     tax, population, live stock and agricul-
     ture, etc. in connexion with the late Ger-
     man New Guinea possessions. 1915.
         Compiled from German official publi-
     cations, etc.

CB   PAPUA AND NEW GUINEA. Dept. of
         Customs and Excise.
         Quarterly bulletin of overseas trade
     statistics.

NEW HEBRIDES

SA   GREAT BRITAIN. Colonial Office.
       Annual report on New Hebrides. 1948-
   Previously issued in its numbered
series, Colonial reports--annual
(325.342   G787) no. 1099: 1920-
no. 1928: 1938. Suspended 1940-46.

     1948-62+   325.342
                                        G7871Nh

SA   FRANCE. Ministère de la France
        d'Outre-Mer.
       Nouvelle-Calédonie et dépendances.
Nouvelles Hébrides.

     HL   1946   DU720
                                          F815

SA   **NEW HEBRIDES.
       Blue book. 1920-(?)

NEW ZEALAND

SA   NEW ZEALAND. Dept. of Island Terri-
        tories.
       Report.

     1951/52-1964/65+   354.931
                                         N532

SA   NEW ZEALAND. Dept. of Statistics.
       Pocket digest of New Zealand statis-
tics.
     Issued through 1936 by the Census and
Statistics Office.
     Title varies slightly.

     1938, 1947-65+   HA3032
                                         A35

SA   NEW ZEALAND. Dept. of Statistics.
       New Zealand official year-book. 1892-
     Issued 1892-1914 by the Registrar
General's Office; 1915-55 by the Census
and Statistics Dept. (1915-36 under an
earlier name: Census and Statistics
Office).

     4th, 1895-1965+   319.31
                                         N532

SA   NEW ZEALAND. Census and Statistics
        Dept.
       Statistics of the Dominion of New Zea-
land.
     In 1 vol. each year, 1853-1902; in 2
vols. each year, 1903-12; in 4 vols. each
year, 1913-20.
     Continued by a series of statistical
reports issued by the Census and Statistics Office, each dealing with a separate
definite branch of statistical inquiry.

     1873, 1896, 1906: v. 1,   319.31
       1907-20   N532s
wanting: 1918: v. 4

SB   NEW ZEALAND. Dept. of Statistics.
       Monthly abstract of statistics. Oct.
1914-
     Issued 1914-21 by the Government
Statistician; 1922-54 by the Census and
Statistics Dept.
     Issues for 1914-16 called v. 1-33.
     Some issues accompanied by supple-
ments.

     Oct. 1914-Mar. 1966+   319.31
wanting: Jan., Mar. 1916,   N531
Jan. 1919, July 1930

CA   NEW ZEALAND. Dept. of Statistics.
       Exports. 1962/63-
     Supersedes in part Statistical report
on the external trade of New Zealand,
formerly issued by the Customs Dept.

     1962/63-1963/64+   HF279
                                           A48

CA   NEW ZEALAND. Dept. of Statistics.
       Imports. 1962/63 (A-B)-
     Supersedes in part Statistical report
on the external trade of New Zealand,
formerly issued by the Customs Dept.
     Issued in two parts: pt. A, Commodity
by country; pt. B, Country by commodity.

     1962/63 (A-B)-1963/64   HF279
       (A-B)+   A481

CA NEW ZEALAND. Customs Dept.
    Statistical report on the external trade of New Zealand. 1950-62//
    Statistics for 1949 included in report for 1952.
    Supersedes pt. 1 of the Statistical report on trade and shipping, issued by the Census and Statistics Dept.
    Issued in two parts: A. Exports; B. Imports.
    Superseded by Exports and Imports, issued by the Dept. of Statistics.

    [1949]-1962   382.93
                                   N532l

CA NEW ZEALAND. Dept. of Statistics.
    Report on and analysis of external trade statistics of New Zealand. 1949-52--1962//
    Began publication in 1954, covering the period 1949-52.
    Supersedes in part its Statistical report on trade and shipping of New Zealand, pt. 2.
    Reports for 1949-52--1953 issued by Census and Statistics Dept.

    1949-52--1962   382.93
                                   N532r

CA NEW ZEALAND. Census and Statistics Dept.
    Statistical report on the trade and shipping of the Dominion of New Zealand. 1921-
    1925- issued in two parts.

    1921--1947-48, pt. 1-2;   382.93
    1950-51   N532s

CA NEW ZEALAND. Census and Statistics Office.
    External trade.

    1948-50   382.93
                                   N532e

CA NEW ZEALAND. Census and Statistics Dept.
    Statistical bulletin.

    no. 3, 1946 (has title Trade   319.31
    and shipping statistics)   N531b

CA NEW ZEALAND. Board of Trade.
    Annual report. no. 1, 1916/17-

    1916/17-1918/19   380.09931
                                   N531

CB NEW ZEALAND. Dept. of Statistics.
    External trade of New Zealand; country analyses. Quarterly.
    Supplement to its Monthly abstract of statistics.

    Bound with its Monthly abstract of statistics (319.31   N531).

AA NEW ZEALAND. Dept. of Statistics.
    Report on the farm production statistics of New Zealand. 1921/22-
    Issued 1921/22-1953/54 by the Census and Statistics Dept. (1921/22-1934/35 under an earlier name: Census and Statistics Office).
    Title varies.
    Supplements accompany some reports.

    1921/22-1963/64+   630.6931
                                   N56

AA NEW ZEALAND. Dept. of Statistics.
    Agricultural and pastoral statistics of New Zealand, 1861-1954.
    Issued as Appendix to Report on the farm production statistics of New Zealand, 1954/55.

    1861-1954 (1 vol.)   630.6931
                                   N56
                                   1954/55
                                   Appendix

AA **NEW ZEALAND. Dept. of Agriculture.
    Statistics. Report of the chief inspector of stock. 1900/01-1908/09//(?)

AB **NEW ZEALAND. Dept. of Agriculture.
    Bulletin. n. s. no. 1, Aug. 1910-

AB **NEW ZEALAND. Dept. of Agriculture.
   Statistical bulletin. Statistics, agricultural and pastoral. 1-5, 1896-1909//(?)
   Each number covers five years, i.e., no. 1 covers 1896/1905; no. 2, 1897/1906, etc.

AR NEW ZEALAND. Dept. of Agriculture.
    Report.

    no. 12, 13, 16-17, 21-22;   630.693
    1903/04-1918/19   N532

NICARAGUA

SA NICARAGUA. Dirección General de Estadística y Censos.
    Resumen estadístico, 1950-1960.

    1950-60 (1 vol.)

SA NICARAGUA. Dirección General de Estadística.
    Anuario estadístico.

    1942-47   HA831
                                       A32

SA GREAT BRITAIN. Dept. of Overseas
CA   Trade.
    ... Economic conditions in the republic of Nicaragua .... Report ....
    [1927-34.]
       1928, Economic conditions in the republic of Nicaragua ... and in the republic of Guatemala ....
       Statistics for 1921/22 and 1930-32 are found in the Report on economic and commercial conditions in the republic of Honduras ..., 1921/22, and in the Report on economic and commercial conditions in the republic of Guatemala ..., 1930-32, respectively.

    1927, 1928, 1934   380.005
                                       G786

SA **NICARAGUA. Ministerio de Hacienda y Crédito Público.
    Estadísticas económicas de Nicaragua.
    1-2, 1904-05//

SB NICARAGUA. Dirección General de Estadística y Censos.
    Boletín de estadística. año 1, July 1944-
    Title varies.

    año 1, no. 3-año 3,   HA831
      no. 41/43; 3d epoca,   A33
      no. 2-10; 1944-64+

CA NICARAGUA. Oficina del Recaudador General de Aduanas.
    Memoria del recaudador general de aduanas y Alta Comisión. 1911/13-
       1911/13-1927 have title: Memoria del recaudador general de aduanas ... y las estadísticas del comercio ....
       Published also in English, 1911/13-1933.

    1918-24, 1929, 1933-64+   382.7285
                                       N583a

    Eng. ed., 1911/13-1933   382.7285
    wanting: 1918, 1923-24   N583

AB NICARAGUA. Ministerio de Agricultura y Ganaderia.
    Boletín. 1959(?)-

AB **NICARAGUA. Ministerio de Agricultura y Trabajo.
    Boletín de agricultura. 1, 1929-

NIGER

SB NIGER. Service de la Statistique.
    Bulletin de statistique. 1, année 1959-

    no. 13-28, Mar. 1962-   HA2117
    Dec. 1965+   N5A4

CA NIGER. Service de la Statistique et de la Mécanographie.
    Commerce extérieur. 1961- Données sur les années antérieures.

    1961+   HF268
                                       N5A3

AR   NIGER. Archives de l'Office du Niger.
     Compte-rendu succinct de la campagne
     agricole.

     FRI 1952/53-1958/59+

AR   NIGER. Service de l'Agriculture.
     Rapport annuel.

NIGERIA

Publications are also listed in subsections as follows:
EASTERN REGION OF NIGERIA
NORTHERN REGION OF NIGERIA
SOUTHERN NIGERIA
WESTERN REGION OF NIGERIA

SA   NIGERIA. Federal Office of Statistics.
     Annual abstract of statistics. 1960-

     1960-61+   HA1977
                           N5A22

SA   NIGERIA. Federal Information Service.
     Federal Nigeria; report. 1957-

     1957-   966.9
                           N686f

SA   GREAT BRITAIN. Colonial Office.
     Annual report on Nigeria. 1946-55//
     Previously issued in its numbered
     series, Colonial reports--annual
     (325.342 G787) no. 1098: 1920-21--
     no. 1904: 1938. Suspended 1940-46.

     1946-54   325.342
                           G7871n

SA   GREAT BRITAIN. Dept. of Overseas
CA   Trade.
     ... Report on economic and commer-
     cial conditions in the British dependen-
     cies in West Africa (the Gambia, Sierra
     Leone, the Gold Coast and Nigeria) ....
     1936/37.

     1936/37   380.005
                           G786

SA   NIGERIA. Chief Secretary's Office.
     Nigeria handbook, containing statisti-
     cal and general information respecting
     the Colony and Protectorate.

     HL 1924, 1927, 1933,   DT515
         1936   N685

SA   NIGERIA.
     Blue book. 1914-38(?)
     Continued the Southern Nigeria Blue
     book (1906-13) and the Northern Nigeria
     Blue book (1900-13). The Southern
     Nigeria Blue book continued the Lagos
     Blue book (1862-1905).

     1916-24, 1926-28   354.669
                           N685

SB   NIGERIA. Federal Office of Statistics.
     Digest of statistics.

     v. 3, no. 1-v. 11, no. 4;   HA1977
     1954-62+   N5A38

CA   NIGERIA. Federal Office of Statistics.
     Review of external trade.
     Issued in its series of Statistical stud-
     ies, FOS.

     1964+   HF266
                           N5A4

CA   NIGERIA. Federal Office of Statistics.
     Trade report.

     1944, 1946-54, 1956-60+   382.669
                                N685

CA   NIGERIA. Board of Customs and Excise.
     Report.

     1927-28, 1930, 1958/59-   HJ7297
     1961/62+   N4A35

CB   NIGERIA. Federal Dept. of Commerce
     and Industries.
     Nigeria trade journal. v. 1, 1953-
     Issued 1953-July/Sept. 1955 by the
     department under its earlier name: Dept.
     of Commerce and Industries.

     v. 1-14, no. 1; 1953-   330.9669
     Mar. 1966+   N689

CB NIGERIA. Federal Office of Statistics.
    Nigeria trade summary.

    Apr. 1935, Oct. 1962

AR NIGERIA. Dept. of Agricultural
      Research.
    Report. 1954/55-

    1954/55-1962/63+   630.9669
                                  N685

    FRI 1960/61-1962/63+

AR NIGERIA. Agricultural Dept.
    Annual report. 1930-1953/54//
    Superseded by Report of the Dept. of
    Agricultural Research, 1954/55-

    1930-1953/54   630.6669
                                  N685r

    FRI 1945-46, 1948-
           1953/54

AR NIGERIA. Agricultural Dept.
    Annual bulletin. 1922(?)-

    2d, 1923; 4th-11th,   630.6669
       1925-32   N685

EASTERN REGION OF NIGERIA

SA EASTERN NIGERIA. Ministry of Eco-
      nomic Planning.
    Eastern Nigeria statistical digest.
    1963-

    1963   HA1977
                                             E2A3

CA EASTERN REGION OF NIGERIA. Minis-
      try of Commerce.
    Report.

    1958/59-1960/61+   338.09669
                                  E15

    HL 1958/59-1960/61+

AR EASTERN REGION OF NIGERIA. Agri-
      culture Division.
    Report.

    FRI 1955/56-1961/62+

NORTHERN REGION OF NIGERIA

SA **NORTHERN NIGERIA.
    Blue book. 1900-13//
    Continued under Nigeria.

AR NORTHERN NIGERIA. Ministry of Agri-
      culture.
    Report.

    1954/55-1963/64, 1962/65+   S338
                                             N57A35

    FRI 1953/54-1962/63+

SOUTHERN NIGERIA

SA **SOUTHERN NIGERIA.
    Annual report.

SA **SOUTHERN NIGERIA.
    Blue book. 1906-13//
    Continues Lagos, Blue book; continued
    under Nigeria.

SA NIGERIA, SOUTHERN. Governor.
    Annual report.... 1899/1900-1913//

    1899/1900-1913   Brit. Docs.

CA **SOUTHERN NIGERIA.
    Trade report of the colony and protec-
    torate of Southern Nigeria.
    Title varies.

AR **SOUTHERN NIGERIA. Agricultural Dept.
    Agricultural bulletin.
    Supplement to the Government gazette.

WESTERN REGION OF NIGERIA

SB WESTERN REGION OF NIGERIA. Ministry of Economic Planning. Statistics Division.
    Western Region statistical bulletin. June 1959-

    v. 1, no. 1-v. 7, no. 2;   316.69
    1959-65+   W527

AR WESTERN REGION OF NIGERIA. Dept. of Agriculture.
    Report.

    1951/52-1954/55   630.6669
                                       W527

NORTH BORNEO

see also BRITISH NORTH BORNEO
         SABAH

SA GREAT BRITAIN. Colonial Office.
    Annual report on North Borneo. 1946-63//
    Previously issued in its numbered series, Colonial reports--annual; suspended 1940.
    1964- to be included in Annual report--Malaysia.
    1963 has title: Sabah, report.

    1947-62   325.342
                                       G7871Nb

CA NORTH BORNEO. Trade and Customs Dept.
    North Borneo statistics, external trade. 1946(?)-

NORTHERN IRELAND

SA NORTHERN IRELAND. Ministry of Finance. Registrar General's Division.
    Ulster year book. 1926-
    1927-28 never published; 1939-46 publication suspended.

    1926, 1929, 1947, 1950,   314.16
    1953, 1956-65+   N874

SB NORTHERN IRELAND. Ministry of Finance.
    Digest of statistics. 1, Mar. 1954-

AA NORTHERN IRELAND. Ministry of Agriculture.
    Annual report upon the agricultural statistics of Northern Ireland. 1st, 1925-

    1925   630.6415
                                       N874

AB **NORTHERN IRELAND. Ministry of Agriculture.
    Monthly report. no. 1, Apr. 1926-

AR NORTHERN IRELAND. Ministry of Agriculture.
    Annual general report. 1, 1921/22-
    Continues in part: Ireland. Dept. of Agriculture and Technical Instruction. Annual general report.

    1921/22-1923/24   S219
                                       A72

NORWAY

SA NORWAY. Statistisk Sentralbyrå.
    Statistisk årbok for kongeriket Norge. 1880-

    1880-1965+   314.81
                                       N892s

SA NORWAY. Statistisk Sentralbyrå.
    Statistisk oversigter, résumé rétrospectif. 1914, 1926, 1948, 1958.
    These volumes contain historical statistics.

    1914, 1926, 1948   314.81
                                       N892st

    FRI 1958

SA **NORWAY. Statistisk Sentralbyrå.
    Norges officielle statistikk. Ser. 1,
1861-80; new Ser., 1883-84; Ser. 3,
no. 1-345, 1885-1900; Ser. 4, no. 1-130,
1901-05; Ser. 5, no. 1-220, 1906-13;
Ser. 6, no. 1-194, 1914-20; Ser. 7,
no. 1-200, 1920-26; Ser. 8, no. 1, 1926-
    Ser. 1-2 not consecutively numbered.
    Subseries and monographs in this
series are catalogued separately.

SA NORWAY. Statistisk Sentralbyrå.
    Annuaire statistique de la Norvège.
no. 1, 1879-84//
    Later issues, see Statistisk årbok.

1879   314.81
                       N892s
                       1879

SA NORWAY. Departementet for det Indre.
    Statistiske tabeller for kongeriget
Norge. 1-20, 1835-59//
    1835-38 issued by Finants-, Handels-
og Tolddepartementet.
    Later volumes issued in Norges offi-
cielle statistikk.

1835, 1838, 1844, 1847,   382.481
1850-59   N892e

SB NORWAY. Statistisk Sentralbyrå.
SA   Statistisk månedshefte. 1, 1882/83-
    Published as Norges officielle sta-
tistikk, 1913-59.
    Title varies: 1882/83-1921, Meddelel-
ser; 1922-49, Statistiske meddelelser;
1950-59, Statistiske meldinger.
    Frequency varies: 1882/83-1913,
1915-22, 1943-44, annual; 1914, 1923-42,
1944- monthly.

v. 1, 1882/83-Feb. 1966+   314.81
                               N892

CA NORWAY. Statistisk Sentralbyrå.
    Utenrikshandel. External trade.
1875-
    Issued in series Norges officielle sta-
tistikk.
    Title varies: 1875-1904, Tabeller
vedkommende Norges handel; 1905-60,
Norges handel.

(continued)

    Issued in parts each year.

1875-77, 1881, 1883-90,   382.481
  1893-98, 1901, 1903-04,   N892h
  1906-10, 1914-64:
  pt. 1-3+
wanting: 1955: pt. 2,
  1962: pt. 2

CA NORWAY. Departementet for det Indre.
    ... Tabeller vedkommende Norges
handel .... 1871-74.
    Issued in series Norges officielle sta-
tistikk.
    Continued by the Statistisk Sentralbyrå.

1871-74//   382.481
                       N892g

CA NORWAY. Departementet for det Indre.
    Tabeller vedkommende Norges handel
og skibsfart. 1860-[1871].
    Issued in series Norges officielle sta-
tistikk.
    Earlier volumes issued in its Sta-
tistiske tabeller for kongeriget Norge.

1861-70   382.481
                       N892f

CA NORWAY. Departementet for det Indre.
    Tabeller vedkommende Norges handel
og skibsfart ....
    Issued as raekke 3, 5, 7, 9-15, 17-20
of Statistiske tabeller for kongeriget
Norge, udgivne efter foranstaltning af
Departementet for det Indre ....
    1835-38 issued by K. Finants-,
Handels- og Told- Departements Foranstaltning.

1835, 1838, 1844, 1847,   382.481
1850-59   N892e

CB NORWAY. Statistisk Sentralbyrå.
    Månedsstatistikk over utrikshan-
delen. Monthly bulletin of external trade.
1913-
    Issued 1913-59 in series Norges offi-
cielle statistikk.
    Issues for 1960- called 48.- Argang.
"Notes" in English.
    Title varies: 1913-59, Maanedsopgaver
over vareomsaetningen med utlandet.

v. 1, 1913-Feb. 1966+   382.481
wanting: several issues   N892

AA   NORWAY. Statistisk Sentralbyrå.
      Jordbruksstatistikk ... (Landbruks-
      areal og husdyrhold m. v.). Superficies
      agricoles et élevage du bétail. Récoltes,
      etc. 1923-
         Title varies: 1923, Representative
      landbrukstelling; 1924-36, Landbruks-
      areal og husdyrhold; 1937- Jordbruks-
      statistikk.

      1939-64+   630.6481
                                  N892jo

      FRI 1923-28, 1930-64+

AA   NORWAY. Fiskeridirektoratet.
      Fiskeristatistikk. Fishery statistics.
      1900-
         Title varies: 1900- Norges fiskerier.

      1900, 1903-06, 1908-14,   338.31
        1916-21, 1924, 1926-35,   N892
        1937, 1938, 1942-63+

AA   NORWAY. Statistisk Sentralbyrå.
      Jordbrug og faedrift. Divers reseigne-
      ments statistiques relatifs à l'agriculture
      et à l'élevage du bétail 1876-
         First report covers period of 10 years:
      1876-85.
         Title varies: 1876-95, Statistiske
      oplysninger vedkommende Norges jord-
      brug og faedrift; 1896-1900, Norges jord-
      brug og faedrift; 1901, Jordbrug og
      faedrift.

      1876/85-1910, 1911/15,   630.6481
        1916/20   N892

AA   NORWAY. Statistisk Sentralbyrå.
      Tabeller vedkommende Norges fiske-
      rier 1879-99//

      1879-99   338.31
                                  N892

AA   NORWAY. Departementet for det Indre.
      Beretninger om Norges fiskerier ....
      1868-
         Issued in series Norges officielle sta-
      tistikk.
         Continued by the Statistisk Sentralbyrå
      with title Tabeller vedkommende Norges
      fiskerier ..., 1879-99, and by the Norges
      Fiskeristyrelse with title Norges fiske-
      rier, 1900-

      1868-78   338.31
                                  N892

NYASALAND

      see also RHODESIA AND NYASALAND
                  (1953-64)
               MALAWI (1964-)

SA   GREAT BRITAIN. Colonial Office.
      Annual report on Nyasaland. 1946-
         Previously issued in its numbered
      series, Colonial reports--annual
      (325.342 G787) no. 1075: 1920-21--
      no. 1902: 1938. Suspended 1940-46.

      1946-62+   325.342
                                  G7871Ny

SA   NYASALAND. Central African Statistical
      Office.
      Statistical handbook of Nyasaland.

      1950, 1952   HA1977
                                  N8A4

SA   GREAT BRITAIN. Dept. of Overseas
CA   Trade.
         ... Report on economic and commer-
      cial conditions in Southern Rhodesia,
      Northern Rhodesia and Nyasaland ....
      [1932-39.]
         Statistics for Northern Rhodesia and
      Nyasaland, 1929/30-1930/32, are found
      in Economic conditions in East Africa and
      in Northern Rhodesia and Nyasaland ...
      Report, 1929/30-1930/32, respectively
      (HC517 E2A3).
         Title varies.

      1932, 1933, 1935, 1939   380.005
                                  G786

SA   NYASALAND.
      Blue book. 1904/05.

CA   NYASALAND. Customs Dept.
     Report on the trade of the Protectorate.
     Ceased publication with the report for 1953.
     Reports for -1927, 1931-32 issued as supplements to the Nyasaland government gazette.
     Continued in Central African Statistical Office. Statement of external trade. (See Rhodesia and Nyasaland.)

     1927, 1932, 1947-53   382.6897
                                 N993

AR   NYASALAND. Agricultural Production and Marketing Board.
     Report. 1, 1956-

AR   NYASALAND. Dept. of Agriculture.
     Annual report.

     FRI 1910/11-1911/12,
          1914/15-1919/20,
          1956-58: pt. I-
          1961/62: pt. II+

P

PAKISTAN

Publications are also listed in subsections as follows:

EAST PAKISTAN
WEST PAKISTAN

SA   PAKISTAN. Central Statistical Office.
     Statistical pocket-book of Pakistan. 1962-

     1962-64+   HA1730.5
                                 S7

     FRI 1963-65+

SA   PAKISTAN. Central Statistical Office.
     Pakistan statistical yearbook.
     Began publication in 1954, covering the period 1952.

     1952, 1955, 1957, 1958,   HA1730.5
     1962+   P33

     FRI 1952, 1957, 1958,
          1962+

SA   PAKISTAN. Dept. of Commercial Intelligence and Statistics.
     Statistical digest of Pakistan. 1950-

     1950 no more published?   315.4291
                                 P152
     FRI 1950

SB   PAKISTAN. Central Statistical Office.
     Statistical bulletin. Mar. 1952-

     no. 10-22; v. 2: no. 2-12; 315.4291
     v. 3-v. 14: no. 4;   P152sb
     Dec. 1952-Apr. 1966+

     FRI Latest 6 months
          only.

CA   PAKISTAN. Central Statistical Office.
CB   Foreign trade statistics of Pakistan.
     Issued quarterly with additional annual cumulation.

     1957-60   HF240.5
                                 A3
     FRI Current two years
          only.

CA   PAKISTAN. Foreign Trade Development Council.
     The foreign trade of Pakistan.

     HL   1951   HF3790.5
                                 P152

AA   PAKISTAN. Dept. of Agricultural Economics and Statistics.
     Fact series.

EAST PAKISTAN

SA   EAST PAKISTAN. Bureau of Statistics.
       Statistical digest of East Pakistan.
     no. 1, 1963-
       Supersedes Statistical abstract for East Pakistan, formerly issued by the Provincial Statistical Board and Bureau of Commercial and Industrial Intelligence.

     1963-64+   HA1730.5
                            E2A3

SA   EAST PAKISTAN. Provincial Statistical Board and Bureau of Commercial and Industrial Intelligence.
       Statistical abstract for East Pakistan.
       Superseded by Statistical digest of East Pakistan, issued by the Bureau of Statistics, 1963-

     v. 3 (1949/50-1953/54),   315.4295
     v. 4 (1952/53-1956/57),   E13
     v. 5 (1950/51-1959/60)+

WEST PAKISTAN

SA   WEST PAKISTAN. Public Relations Dept.
       West Pakistan yearbook. 1956-

     1958, 1963   315.4291
                             W522

SA   WEST PAKISTAN. Bureau of Statistics.
       Statistical handbook of West Pakistan.

AA   WEST PAKISTAN. Food Directorate of West Pakistan.
       Food and agricultural statistics of West Pakistan.

     FRI   1959

PALESTINE

see also ISRAEL, 1948-

SA   PALESTINE.
       Statistical handbook of Jewish Palestine. 1947-

     1947   315.69
                             J59s

SA   PALESTINE.
       Blue book. 1926/27-

     1926-31, 1935-36, 1938   354.569
                                   P157

SA   GREAT BRITAIN.
       Report by His Britannic Majesty's government to the Council of the League of Nations on the administration of Palestine and Transjordan. 1920/21-
     1923- in Gt. Brit. Colonial Office. Colonial series. 1920/21-1923 as Government of Palestine administration. (Other changes of title.)

     1920-22   325.342
                               G784
                               no. ob

     HL   1920-36   DS125
                               A6G783

SA   GREAT BRITAIN. Dept. of Overseas
CA   Trade.
       ... Report ... Economic conditions in Palestine .... [1927-35.]
       Title varies.

     1927, 1931, 1935   380.005
                               G786

SB   PALESTINE. Dept. of Statistics.
       General monthly bulletin.
       Includes Palestine commercial bulletin (HF259   P3A25).

     1922-33, 1947-48   HA1931
     wanting: many issues   A3
                               f

CA   PALESTINE. Dept. of Statistics.
     Statistics of foreign trade.
     1925-27 issued by Dept. of Customs,
     Excise and Trade.
     Title varies.

     1925-27   HF259
                                      P3A26
     1941, 1942/43, 1944/45   HF259
                                      P3A3

CA **PALESTINE. Dept. of Customs, Excise
     and Trade.
     Annual statistical return of the government of Palestine customs. no. 1-2,
     1919/20-1920/21//(?)
     No. 1 as Annual statistical return of
     the O.E.T.A. South, customs.

AR   PALESTINE. Dept. of Agriculture and
     Forests.
     Report. 1924-

     1927-30   630.6569
                                      P157

PANAMA

SA   PANAMA. Dirección de Estadística y
     Censo.
     Panamá en cifras; compendio estadística, años ....

     1958-62 (1 vol.),   317.287
       1959-63 (1 vol.),   P188p
       1960-64 (1 vol.)

SA   PANAMA. Dirección de Estadística y
     Censo.
     Extracto estadístico. Estadística
     general. 1941/43-

     1944-49, 1950-52,   HA851
       1953-54   A385

SA   GREAT BRITAIN. Dept. of Overseas
CA   Trade.
     ... Report on economic and commercial conditions in the republic of Panama
     and the Panama Canal Zone ... and in the
     republic of Costa Rica .... [1921-37.]
     Statistics for Panama for 1925 are
     found in Report on the economic, financial and commercial conditions in the
     republics of Costa Rica ... and Panama
     ..., 1925.
     Title varies.

     1921, 1922, 1924, 1927,   380.005
       1929, 1931-37   G786

SA **PANAMA. Dirección General de Estadística.
     Boletín de estadística. Anuario.
     1911//
     Continues Estadística anual. Continued as Boletín estadística del quinquenio.

SA **PANAMA. Dirección General de Estadística.
     Estadística anual. 1908-10//
     Continued as Boletín de estadística.
     Anuario.

SB   PANAMA. Dirección de Estadística y
     Censo.
     Estadística panameña. Oct. 1941-
     v. 17, 1958//
     Issued by the Controlería General de
     la República, Sección Estadística, Oct.
     1941-Mar. 1943; by the Dirección General de Estadística, Apr. 1943-Dec. 1945.
     Continued in subseries.

     v. 1, no. 4-v. 17; 1941-58   317.287
     wanting: v. 1, no. 12,   P188
       Sept. 1942; no. 2, 1958

SB   PANAMA. Dirección General de Estadística.
     Boletín de estadística. Publicación
     oficial. 1, June 1907-

     no. 29^2-50, 1915-22   317.287
     wanting: no. 30^1, 31-33,   P187
       36-37: 1917

SB **PANAMA. Secretaría de Fomento y
AB   Obras Públicas.
      Boletín estadístico. Publicación trimestral.
      Superseded by Boletín de estadística de la República de Panamá.

CA   PANAMA. Dirección de Estadística y Censo.
      Estadística panameña. Serie K.1. Anuario de comercio exterior. 1958-

    1958-63+   317.287
                              P1881
                              Ser. K.1

CA   PANAMA. Dirección de Estadística y Censo.
      Anuario de comercio exterior. (?) -1957//
      Continued in its Estadística panameña. Serie K.1. Anuario de comercio exterior. 1958-

    1953-57   382.7287
                              P187a

CA   PANAMA. Dirección de Estadística y Censo.
      Extracto estadístico de comercio exterior. 1941-43--(?)

    1941-43, 1944-46,   382.7287
    1947-48, 1951/52   P187

CB   PANAMA. Dirección de Estadística y Censo.
      Estadística panameña. Serie K. Comercio exterior.
      Began publication with v. 18, Mar. 1959, in continuation of the volume numbering of its Estadística panameña, Oct. 1941-58, which it supersedes in part.

    1958-64, no. 2+   317.287
                              P1881
                              Ser. K

AA   PANAMA. Dirección de Estadística y Censo.
      Estadística panameña. Serie H. Información agropecuaria.
      Began publication with v. 18, Apr. 1959, in continuation of the numbering of its Estadística panameña, Oct. 1941-58, which it supersedes in part.
      Issued in 4 (no. 1-4) subseries.

    no. 1: 1958, 1961/62,   317.287
           1963/64, 1964/65   P1881
    no. 2: 1958/59, 1961/62,   Ser. H
           1962/63, 1963/64,
           1964/65
    no. 3: 1962, 1964
    no. 4: 1959, 1962

AA   PANAMA. Dirección de Estadística y Censo.
      Información agropecuaria. v. 1-3, 1956-58.
      Continued in its Estadística panameña. Serie H. Información agropecuaria.

AA **PANAMA. Dirección General de Estadística.
      Boletín estadístico del quinquenio ... Sección agrícola y varios. 1912/16//

AB   PANAMA. Dirección de Estadística y Censo.
      Estadística panameña. Serie H.1. Información agropecuaria. Precios recibidos por el agricultor. 1958-
      Began publication with v. 18, Feb. 1959, in continuation of the numbering of its Estadística panameña, Oct. 1941-58, which it supersedes in part.

    v. 18-25: 2, 1959-Feb.   317.287
       1966+   P1881
                              Ser. H.1

AB **PANAMA. Secretaría de Agricultura y Obras Públicas.
      Boletín agrícola. 1, Sept. 1925-

AR   PANAMA. Secretaría de Agricultura y Obras Públicas.
      Memoria. 1924/26-

PAPUA

SA   AUSTRALIA. Dept. of Territories.
        Territory of Papua: annual report.
        Title varies.

        1948/49-1957/58 (for later   354.995
          issues see bound set of   A938p
          Parliamentary Papers
          (328.94 A938r)

SA   PAPUA. Lieutenant-Governor.
        Annual report for the year.

        1920/21   J964
                                   A7
                                   f

SA   **PAPUA.
        Bulletin of the Territory of Papua.
        no. 1-7, 1913-21//
        Issued by Australia. Dept. of Home
    and Territories.

SA   GREAT BRITAIN.
        Reports on British New Guinea.
    1889/90-

    1899/1900   Brit. Docs.

PAPUA-NEW GUINEA (Territory)

SB   PAPUA-NEW GUINEA (TERRITORY).
        Bureau of Statistics.
        Summary of statistics. no. 1, June
    1959- Quarterly.

    no. 12-25, Mar./June   HA4007
        1962-Sept. 1965+   P3A3
    wanting: no. 23   f

CA   PAPUA-NEW GUINEA (TERRITORY).
        Bureau of Statistics.
        Oversea trade. Bulletin.
        Vols. for 1955/56-1957/58 issued by
    Dept. of Customs and Marine.

    1963/64-1964/65+   HF293
                                    P3A34

CB   PAPUA-NEW GUINEA (TERRITORY).
        Bureau of Statistics.
        Bulletin of oversea trade statistics.

    Sept. 1964-Sept. 1965+   HF293
                                    P3A35

AR   PAPUA-NEW GUINEA (TERRITORY).
        Dept. of Agriculture, Stock and Fish-
    eries.
        Report. 1, 1959/60-

    FRI 1959/60, 1960/61,
              1961/63

PARAGUAY

SA   PARAGUAY. Dirección General de Esta-
        dística y Censos.
        Anuario estadístico de la República del
    Paraguay. 1886-
        Agency varies.
        Publication suspended, 1888-1913 and
    1918-39, inclusive. Statistical informa-
    tion for the latter period is contained in
    the annual reports of the Ministry of
    Finance.

    1915, 1942-47, 1948-53,   HA1041
        1954-59+   A2

    FRI 1958-59+

SA   PARAGUAY. Ministerio de Economía.
        Datos y cifras estadísticas: población,
    producción, importación, exportación,
    industrias, vialidad, comercio, instruc-
    ción pública. 1939.

    1939

SA   PARAGUAY. Dirección General de Esta-
        dística y Censos.
        Memoria ....

    1934, 1936-37, 1938   318.44
                                    P222m

SA GREAT BRITAIN. Dept. of Overseas
CA   Trade.
       ... Report on economic and commercial conditions in Paraguay.... [1921-36.]
       Title varies.

       1921, 1922, 1923, 1924,   380.005
       1933, 1936   G786

SB PARAGUAY. Dirección General de Estadística y Censos.
       Boletín estadístico del Paraguay. 1, 1957-

       v. 2, no. 6-v. 7, no. 19/21;   318.44
       June 1958-Dec. 1964+   P222b
       wanting: 1963-Nov. 1964

SB **PARAGUAY. Dirección General de Estadística.
       Boletín trimestral. no. 1, 1915-

SB **PARAGUAY. Dirección General de Estadística.
       Boletín trimestral. 1-2 (no. 1-8), 1906-07//

CA PARAGUAY. Dirección General de Estadística y Censos.
       Importación y exportación de acuerdo a la nomenclatura de Bruselas. Año ....

       1920   382.89
                                    P222

CA PARAGUAY. Dirección General de Estadística y Censos.
       Estadísticas de importación y exportación de la República del Paraguay en ....

       1919   382.89
                                    P222e

PERSIA

see IRAN

PERSIAN GULF

see BAHREIN

PERU

SA PERU. Dirección de Estadística.
       Anuario estadístico del Perú. 1944/45-
       Continues its Extracto estadístico del Perú.

       1944/45-1956/57   HA1051
                                    A32

       FRI 1954/55

SA PERU. Dirección de Estadística.
       Extracto estadístico del Perú.

       1919-20, 1925-28, 1931-43   318.5
       wanting: 1939/40   P471

SA **PERU. Dirección General de Estadística.
       Statistical abstract of Peru. 1918-
       1921-22 never published.
       Spanish edition. See its Extracto estadístico del Perú.

SB PERU. Dirección Nacional de Estadística y Censos.
       Boletín de estadística peruana. 1, 1958-
       Supersedes Boletín de estadística peruana issued by the Dirección de Estadística.

       no. 5-7, 1961-64+   HA1051
                                    A33

SB PERU. Dirección de Estadística.
       Boletín de estadística peruana. no. 1, June 1929-

       v. 7-16: no. 2, 1946-62   318.5
       wanting: v. 9, no. 2   P471b

CA   PERU. Superintendencia General de
CB   Aduanas.
       Anuario del comercio exterior del
Perú ... Publicación oficial.
      Title and frequency vary: Sept. 1922-
Dec. 1926, Estadística del comercio
exterior; Jan. 1927-June 1930, Comercio
exterior. Estadística de enero [etc.];
1931- Anuario ....
      Sept. 1922-Dec. 1923, monthly;
1924-30, quarterly; 1931- annual.

1922-49, 1951-54,   382.85
  1958-63+   P471e
wanting: 2d and 3d semes-
  ter 1930

CA   PERU. Superintendencia General de
CB   Aduanas.
      Estadística del comercio especial del
Perú en el año .... 1919-Dec. 1936.
      1933 has title: Boletín semestral del
comercio especial del Perú...; 1934-35,
Boletín trimestral del comercio especial
del Perú; 1936, Boletín mensual del
comercio especial del Perú.
      Discontinued with Dec. 1936 issue.

1917, 1919, 1921-30,   382.85
  1932-36   P471

CA   PERU. Superintendencia General de
       Aduanas.
      Estadísticas de importación y exporta-
ción de la República del Perú en el año
... conforme a la nomenclatura de
Bruselas de 1913 y con los valores
expersados en libras peruanas y pan-
americanos.

1921-22   382.85
                                P472

CA **PERU. Superintendencia General de
       Aduanas.
      Estadística del comercio especial.
1902-12(?)//
      1902-09 has supplementary volume
with title: Leyenda de la nomenclatura
comercial de la estadística. (Title
varies.) Continues Ministerio de
Hacienda. Estadística general de
aduanas.

CA **PERU. Superintendencia General de
       Aduanas.
      Estadística general de aduanas. 1891-
1901//
      Issued in two parts each year: Expor-
tación, cabotaje; Importación.
      Continued as Superintendencia General
de Aduanas. Estadística del comercio
especial.

CA   PERU. Ministerio de Hacienda y Comer-
       cio. Dirección General de Comercio.
      Informaciónes comerciales.

CB **PERU. Superintendencia General de
       Aduanas.
      Boletín de estadística comercial. 1-5
(no. 1-62), Nov. 1903-July 1908//

AA   PERU. Ministerio de Agricultura.
      Estadística agraria. Agricultural sta-
tistics. 1963(?)-

FRI   1963

AA   PERU. Dirección de Economía Agraria.
      Estadística agropecuaria. 1962.

AA   PERU. Ministerio de Agricultura. Ser-
       vicio Cooperativo Interamericano de
       Producción de Alimentos (SCIPA).
       División de Estúdios Económicos.
      La situación agropecuaria en el Perú
1946 a 1956.

FRI   1946-56 (1 vol.)

AA   PERU. Ministerio de Agricultura.
       Dirección de Economía Agropecuaria.
      Anuario monográfico agropecuario.

FRI   1953

AA   PERU. Dirección de Economía Agro-
       pecuaria.
      Resumen estadístico de la producción
agropecuaria del pais.
      Name of issuing office varies:
Departamento de Estadística Agropecuaria.

PHILIPPINE ISLANDS

see also PHILIPPINES (COMMONWEALTH)
PHILIPPINES (REPUBLIC)

SA   GREAT BRITAIN. Dept. of Overseas
CA   Trade.
       ... Trade conditions in the Philippine
Islands ... Report .... [1925-38.]
       Title varies.

      1925-30, 1932-34, 1938   380.005
                                                 G786

SA   PHILIPPINE ISLANDS
      Guia oficial de las Islas Filipinas.
1834(?)-98//
      1934 as Almanaque filipino i guia de
forasteras; 1879-81 as Guia de forasteras en Filipinas ... Anuario estadístico.

      1897   354.914
                                          P552

SB   PHILIPPINE ISLANDS. Dept. of Agri-
      culture and Commerce. Division of
      Statistics.
      Bulletin of Philippine statistics. 1934-
      Title varies: v. 1-4, 1934-37, The
Philippine statistical review.

      v. 1-6: no. 3, 1934-   319.14
      Sept. 1939   P552

CA   PHILIPPINE ISLANDS. Bureau of Com-
      merce.
      Report.
      Ceased publication with the report for
1934.
      Continued by Philippines (Common-
wealth). Bureau of Commerce. Report.

      1934 (tables have statistics   382.914
      1899-1934)   P551r

CA **PHILIPPINE ISLANDS. Intendencia
      General de Hacienda.
      Estadística general del comercio
exterior de las Islas Filipinas....
      Continues Balanza mercantil.
      Title varies slightly.

CA **PHILIPPINE ISLANDS. Intendencia
      General de Hacienda.
      Balanza mercantil de las Filipinas....
1857-61//
      Continued as Estadística general del
comercio exterior de las Islas Filipinas.

CA **PHILIPPINE ISLANDS. Intendencia
      General de Hacienda.
      Cuadro general del comercio exterior.
1856//
      Continued as Balanza mercantil ....

CB   PHILIPPINE ISLANDS. Bureau of Com-
      merce.
      Statistical bulletin of the Philippine
Islands. 1918-
      No. 1-12 issued by the bureau under
its earlier name: Bureau of Commerce
and Industry.
      Title varies slightly.

      no. 1-12, 1918-29   319.14
                                   P548

AR   PHILIPPINE ISLANDS. Bureau of Agri-
      culture.
      Report.
      Report for 1906/07-1912/13, 1917,
and 1921 issued in Philippine agricultural
review (630.691   P552).

      1919-28   630.9914
                                   P551r

PHILIPPINES
(Commonwealth)

SA   PHILLIPINES (COMMONWEALTH).
      Dept. of Agriculture and Commerce.
      Facts and figures about the Philip-
pines. 1939.

      1939   919.14
                                   P555

CA   PHILIPPINES (COMMONWEALTH).
     Bureau of Commerce.
     Report. 1935(?)-
     Continues the Report of the Bureau of Commerce of the Philippine Islands.

     1936-37, 1939 (Jan.-   382.914
       June), 1939/40   P552r

CA   PHILIPPINES (COMMONWEALTH).
     Bureau of Customs.
     Report of the insular collector of customs.
     Reports for 1910/11-1934 issued by the Bureau of Customs of the Philippine Islands.
     Report for July 1913/Dec. 1914 has title: Foreign commerce of the Philippine Islands.

     1911, 1913-38   336.26
                                          P552

PHILIPPINES (Republic)

SA   PHILIPPINES (REPUBLIC). Bureau of the Census and Statistics.
     Facts and figures about economic and social conditions of the Philippines. 1946-47--

     1946-47, 1948-49, 1963   330.9914
                                        P555

SA   PHILIPPINES (REPUBLIC). Bureau of the Census and Statistics.
     Statistical handbook of the Philippines.
     Consists of condensed summaries of statistical data.

     1932, 1952, 1903/53   HA1821
       (1 vol.), 1962+   A3

SA   PHILIPPINES (REPUBLIC). Bureau of the Census and Statistics.
     Yearbook of Philippine statistics. 1940-

     1940, 1946, 1958   319.14
                                      P551

SB   PHILIPPINES (REPUBLIC). Bureau of the Census and Statistics.
     Journal of Philippine statistics. v. 1, July 1941-
     Publication suspended -Dec. 1948.
     Issued July 1941- by the Bureau of the Census and Statistics of the Commonwealth of the Philippines.

     Jan.-June 1949, Apr.-Dec.   HA1821
       1954, Jan.-Sept. 1955,   J68
       Apr.-Dec. 1959, 1960,   f
       June-Dec. 1961, Apr.-
       Dec. 1962, Jan.-May,
       July-Dec. 1963+

     FRI   Current two years only.

SB   PHILIPPINES (REPUBLIC). Bureau of the Census and Statistics.
     Abstract of Philippine statistics. v. 1, no. 1-4; Oct./Dec. 1950-July/Dec. 1951//

     v. 1, no. 1-4   319.14
                                       P551a

CA   PHILIPPINES (REPUBLIC). Dept. of Commerce and Industry.
     Résumé, foreign trade statistics of the Philippines.
     Before 1963, issued by Bureau of the Census and Statistics.

     1963+   HF249
                                         A45

CA   PHILIPPINES (REPUBLIC). Bureau of the Census and Statistics.
     Foreign trade statistics of the Philippines. Imports and exports.
     Title varies: -1949, Foreign trade statistics of the Philippines; 1950- Foreign trade and navigation of the Philippines.

     1945-52, 1963+   382.914
                                       P5529

CB PHILIPPINES (REPUBLIC). Bureau of
    the Census and Statistics.
    Monthly report on foreign trade sta-
tistics. 1961-
       Title varies: Jan.-June 1961, For-
eign trade statistics of the Philippines;
July-Dec. 1961, Monthly foreign trade
statistics report.

    Jan. 1962-June 1965+   HF249
                                  A65

AA PHILIPPINES (REPUBLIC). Dept. of
    Agriculture and Natural Resources.
    Division of Agricultural Economics.
    Handbook of agriculture.

    FRI 1955, 1959

AA PHILIPPINES (REPUBLIC). Dept. of
    Agriculture and Natural Resources.
    Division of Agricultural Economics.
    Philippine agricultural statistics.
    v. 1, 1954; v. 2, 1956//

    1954   630.6914
                                  P554s
    FRI 1954

AB PHILIPPINES (REPUBLIC). Dept. of
    Agriculture and Natural Resources.
    Division of Agricultural Economics.
    Philippine agricultural situation.

    FRI 1959, 1960

AR PHILIPPINES (REPUBLIC). Dept. of
    Agriculture and Natural Resources.
    Report.
       Issued by Philippine Islands. Dept. of
    Agriculture and Natural Resources,
    1930-32; by Philippine Islands. Dept. of
    Agriculture and Commerce, 1933-35; by
    Philippines (Commonwealth). Dept. of
    Agriculture and Commerce, 1936-

    1930-38, 1948/49,   630.9914
       1958/59, 1960/61-   P552r
       1962/63+

AA PHILIPPINES (REPUBLIC). Dept. of
    Agriculture and Natural Resources.
    Division of Agricultural Economics.
    Agricultural and natural resource sta-
tistics. 1958-

AA PHILIPPINES (REPUBLIC). Dept. of
    Agriculture and Natural Resources.
    Division of Agricultural Economics.
    Crop, livestock and natural resources
statistics. 1954-55(?)--

POLAND

   Detailed statistical data on specific fields
(industry, etc.) of the national economy may
be found in the comprehensive publications in
the series Statystyka Polski (Polish statistics).

SA POLAND. Główny Urzad Statystyczny.
    Statystyka Polski. 1, 1919-

    HL [1919-33]   Uncat. coll.

        zesz. 33, 62, 64, 80+   HA1451
                                 A48

SA POLAND. Główny Urzad Statystyczny.
    Concise statistical year book of the
Polish People's Republic. 1930-
       Published 1930-38 by the Chief Bureau
    of Statistics; Sept. 1939/June 1941 by the
    Polish Ministry of Information (of the
    Polish Government in exile); 1959-   by
    the Central Statistical Office.
       Translation of its Mały rocznik sta-
tystyczny.

    1931-38, 1939/41,   HA1451
       1959-65+   A5

    HL 1930-1939/41   HA1451
                                  A42

        1959-65+   HA1451
                                  A32

    FRI 1930-38, 1959,
           1961, 1963-65+

SA POLAND. Główny Urzad Statystyczny.
    Mały rocznik statystyczny. 1958-
       For English edition see its Concise sta-
tistical year book of the Polish People's
Republic.

    HL 1959-65+   HA1451
                                  A3

SA POLAND. Główny Urzad Statystyczny.
    Rocznik statystyczny.
    Began publication in 1930; suspended
    1939-44.
    Published also in English, French,
    and German.

    1949-50, 1955-65+   HA1451
                               A46

HL 1934-39, 1947-50,   HA1451
    1955-63+   A46
    wanting: 1960

FRI Current two years
    only.

SA POLAND. Główny Urzad Statystyczny.
    Statisticheskii ezhêgodnik; perevod.
    Rocznik statystyczny. Statistical year-
    book. 1957-
    Russian, Polish, and English; consists
    of the text part, including preface, chap-
    ter, and column headings of the Rocznik
    statystyczny.

    1957-58   HA1451
                               A466

SA POLAND. Główny Urzad Statystyczny.
    Polska w liczbach.

HL 1944-61, 1944-64   HA1451
                               A47

FRI 1944-61, 1944-64

SA POLAND. Główny Urzad Statystyczny.
    Poland in figures.

    1944-58 (1 vol.)   314.38
                               P762p

HL 1944-58   HA1451
                               A472
                               1959

FRI 1944-64

SA POLAND. Główny Urzad Statystyczny.
    Statistical year book of Poland.
    Began publication with the volume for
    1930; suspended July 1941-46.
    Translation of its Rocznik statysty-
    czny.
    No longer published.

    1947-49   314.38
                               P762ra

HL 1948-49   HA1451
                               A42a

SA POLAND. Główny Urzad Statystyczny.
    Petit annuaire statistique de la
    Pologne.
    Translation of Mały rocznik statysty-
    czny.
    Issued also in English and German.

HL 1932-39   HA1451
                               A44

SA GREAT BRITAIN. Dept. of Overseas
CA   Trade.
    ... Economic conditions in Poland ...
    Report .... [1921-38.]
    Title varies.

    1921-31, 1934-38   380.005
                               G786

SA POLAND. Główny Urzad Statystyczny.
    Rocznik statystyki, Rzeczypospolitej
    Polskiej. Annuaire statistique de la
    République Polonaise. 1920/21-

HL 1920/21-1930   HA1451
    1923, 1924 in   A25
    French only

SA **POLAND. Główny Urzad Statystyczny.
    Les nouvelles statistiques. 1-2,
    1923-24//

SA POLAND. Główny Urzad Statystyczny.
    Statistique de la Pologne.

HL Scattered holdings,
    1919-38 in main
    series, Ser. A, B, C

SA   POLAND. Główny Urzad Statystyczny.
Przyczynki do statystyki byłego
krolestwa Polskiego .... Contributions
à la statistique de l'ancien royaume de
Pologne. 1-3, 1918-20//

     HL   1918

SB   POLAND. Główny Urzad Statystyczny.
Biuletyn statystyczny. Monthly. 1,
Jan. 1957-

     HL   Jan. 1957-Dec. 1965+

     FRI   Current two years only.

SB   POLAND. Główny Urzad Statystyczny.
Wiadomości statystyczne. Informations statistiques.

     HL   1923-39, 1945-50,
           1958-63+
       wanting: several issues

     FRI   Current two years only.

SB   POLAND. Główny Urzad Statystyczny.
Kwartalnik statystyczny. Revue trimestrielle de statistique 1, 1924-
Continued Sprawozdania miesieczne z
handlu zagranicznego. Rapports mensuels du commerce étranger, and
Miesiecznik statystyczny ... Revue
mensuelle de statistique ....

     HL   v. 1-11, 1924-34

SB   POLAND. Główny Urzad Statystyczny.
Miesiecziny statystyczny wydawany
prez Główny Urzad Statystyczny Rzeczypospolitij Polskiej ... Rok .... Revue
mensuelle de statistique publiée par
l'Office Central de Statistique de la
République Polonaise. Année ....
Continued in two publications: Kwartalnik statystyczny and Wiadomości statystyczny.

     HL   v. 1-6: 7-12, 1920-23

CA   POLAND. Główny Urzad Statystyczny.
Statystyka handlu zagranicznego;
import i eksport towarów według krajów.

     HL   1959, 1961 (its series,   HA1451
          Statystyka Polski,   A48
          33, 62)

     FRI   1959, 1960, 1961 (its
           series, Statystyka
           Polski, 33, 53, 62)

CA   POLAND. Główny Urzad Statystyczny.
Statystyka handlu zagranicznego,
obroty handlowe Polski z poszczegolnyme
krajami. 1960.
     Issued as its Statystyka Polski,
zeszyt, no. 56.

     HL   1960

     FRI   1960

CA   POLAND. Główny Urzad Statystyczny.
Statystyka Polski.
Rocznik handlu zagranicznego Rzeczypospolitej Polskiej i Wolnego Miasta
Gdańska. Annuaire du commerce extérieur de la République Polonaise et de la
Ville Libre de Dantzig.
     In Polish and French.
     Published as series since 1931; vol.
for 1931 and 1932 as Seria A, the following as Seria C.
     Title varies slightly.

     HL   1922-25, 1926-27, 1928,
           1929, 1929-30,
           1931-37, 1938

CA **POLAND. Główny Urzad Statystyczny.
Przywoz i wywoz .... Importation et
exportation. 1921-22//(?)

CB   POLAND. Główny Urzad Statystyczny.
Handel zagraniczny Rzeczypospolitej
Polskiej. Commerce extérieur de la
République Polonaise et de la V.L. de
Dantzig.
     In Polish and French. In 1922-24
there were separate Polish and French
editions.
     In 1922-26 published as quarterly,
later as monthly.

     HL   1925-29, 1934-May 1939

CB   POLAND. Główny Urzad Statystyczny.
      Sprawozdania miesieczne z handlu
      zagranicznego.
      Title and text in Polish and French.

AA   POLAND. Główny Urzad Statystyczny.
      Statystyka rolnicza. Statistique agricole.
      In Polish and French.
      Vol. for 1930-32 published as Seria B;
      later as Seria C.

      FRI   1930/31, 1931/32,
            1933-38, 1947

PORTUGAL

SA   PORTUGAL. Instituto Nacional de Estatística.
      Anuário estatístico.
      Issuing body varies: 1875-1887/92,
      Direcção Geral do Comércio e Indústria;
      1893-1934, Direcção Geral da Estatística.
      Vols. for 1961- issued in 2 vols.:
      v. 1, Metrópole; v. 2, Ultramar. Vol. 2
      was formerly issued separately as
      Anuário estatístico do ultramar.

      1930-32, 1934-38, 1942-44,   HA1571
         1947-53, 1955-64: v. 1+   A285

      FRI   1961-63: v. 2+

SA   GREAT BRITAIN. Dept. of Overseas
CA   Trade.
      ... Economic conditions in Portugal
      ... Report .... [1924-36.]
      1926/28 include also notes on the
      financial situation in Angola and Mozambique.
      1930 includes also reports on Madeira
      and Cape Verde islands.
      Title varies.

      1924, 1926, 1928, 1932,   380.005
         1934, 1936   G786

SB   PORTUGAL. Instituto Nacional de Estatística.
      Boletim: ultramar. Outre-mer.

      1951-64+   HA1577
                                 A3

      FRI   Current two years
            only.

SB   PORTUGAL. Instituto Nacional de Estatística.
      Boletim mensal. Bulletin mensuel.
      Jan. 1929-
      Title varies: Boletim mensal da
      Direcção Geral da Estatística (Jan. 1929-
      Apr. 1935).

      1947-65+   314.69
      wanting: Sept. 1956, Jan. P85b
         1957, 1961

      FRI   Current two years
            only.

CA   PORTUGAL. Instituto Nacional de Estatística.
      Comércio externo. 1938-

      1941, 1950: v. 1-2--1964: 380.09469
         v. 1-2+   P853

      FRI   Current two years
            only.

CA **PORTUGAL. Direcção Geral da Estatística.
      Estatística comercial.
      1897-1912 as Comércio e navegação.
      Estatística especial; 1913-20, Estatística
      comercial. Comércio e navegação.

CB **PORTUGAL. Direcção Geral da Estatística.
      Boletim comercial e maritimo.
      Comercio com os paises estrangeiros e
      colónias portuguesas.
      Title varies.

AA   PORTUGAL. Instituto Nacional de Estatística.
      Estatística agrícola.

      1943-52   630.6469
                                 P85

      FRI   1943-64+
            wanting: 1957

AA **PORTUGAL. Direcção Geral da Estatística. Estatística Agrícola.
      Resumos estatísticos. no. 1-4,
      1912-13//

AB **PORTUGAL. Direcção Geral da Agricultura.
   Boletim. ano 1-13; no. 6, 1889-1915//
   Includes agricultural legislation, which for 1889-90 forms a separate volume.

AR **PORTUGAL. Direcção Geral da Agricultura.
   Jornal oficial de agricultura. 1-4, July 1877-June 1881//

PORTUGAL, COLONIES

SA PORTUGAL. Instituto Nacional de Estatística.
    Anuário estatístico do ultramar. Annuaire statistique d'outre-mer.
    Title varies: -1949, Anuário estatístico do império colonial. Annuaire statistique de l'empire colonial.
    Ceased publication with the volume for 1960.
    Beginning with 1961, this material is included in the institute's Anuário estatístico, as v. 2, Ultramar.

    1935-46, 1949-60   314.69
                              A636

    FRI   1950-60

SA **PORTUGAL. Ministério das Colónias.
    Anuário colonial 1916-
    Issued 1916, 1917/18, 1925/27, 1927/29.

SA **PORTUGAL. Ministério das Colónias.
    Estatística geral das colónias. 1-2, 1915//

PORTUGUESE EAST AFRICA

see also MOZAMBIQUE

SA   GREAT BRITAIN. Dept. of Overseas
CA   Trade.
    ... Report on economic and commercial conditions in Portuguese East Africa ....
    Title varies.

    1921, 1927, 1929, 1932,   380.005
    1935, 1938   G786

PORTUGUESE GUINEA

see GUINEA, PORTUGUESE

PORTUGUESE INDIA

SA INDIA, PORTUGUESE. Repartição Central de Estatística e Informação.
    Anuário estatístico. 1932-34, 1948-
    Portuguese and French.
    None published 1935-47.

    1951   HA1728
                              P6A3

SA INDIA, PORTUGUESE. Repartição da Estatística.
    Anuário da India Portuguesa. 1929-
    Title varies.

    ano 5, 1934/35; ano 6,   354.5439
    1936/39   A63

SB INDIA, PORTUGUESE. Repartição Central de Estatística e Informação.
    Boletim trimestral. Jan.-Mar. 1948-

CA **INDIA, PORTUGUESE. Direcção dos Serviços Aduaneiros.
    Estatística, comércio e navegação ....
    1916/20-

CA **INDIA, PORTUGUESE.
    Estatística do comércio de estado da India. 1916-

CA **INDIA, PORTUGUESE.
    Estatística do comércio e navegação da India, Macau e Timor ... e resumos do movimento comercial 1901/03//

CA INDIA. Commercial Intelligence and
       Statistics Dept.
    Accounts of the trade of the Portuguese possessions in India. 1899/1900-
    1875/76-1898/99, 1904/05-1913/14 in
    Annual statement of the trade of British
    India with foreign countries. 1914/15-
    as a supplement. 1916/20 is for a five-
    year period (see India CA).

    1916/20   382.54
                           I38s

AB **INDIA, PORTUGUESE. Direcção dos
       Serviços Agrícolas e Florestals.
    Boletim do fomento. 1-3, 1919-21//
    1-2 as Boletim de agricultura.

AB **INDIA, PORTUGUESE.
    Boletim das alfandegas. 1-9, 1911/13-
    1919//(?)

R

REUNION

SA FRANCE. Institut National de la Statistique et des Etudes Economiques pour
    la Métropole et la France d'Outre-Mer.
    Annuaire statistique de la Réunion.

    1952-55, 1955-58, 1958-60   316.981
                                F815

SA **REUNION ISLAND.
    Annuaire de l'Ile de la Réunion. 1852-
    Continues Almanach de l'Ile Bourbon.
    At head of title: République Française.

SA **REUNION ISLAND.
    Almanach de l'Ile Bourbon. 1848//
    Continued as Annuaire de l'Ile de la
    Réunion.

SB **REUNION ISLAND.
    Bulletin officiel de l'Ile de la Réunion.
    1848-1909//
    Continues Bulletin officiel de l'Ile
    Bourbon.

AA **REUNION ISLAND. Chambre d'Agriculture.
    Bulletin. 1876-

RHODESIA

Publications are listed as follows:

    RHODESIA (NORTHERN AND SOUTHERN)
    RHODESIA, NORTHERN
    RHODESIA, SOUTHERN
    RHODESIA AND NYASALAND
    RHODESIA (1964-)

see also MALAWI
          NYASALAND
          ZAMBIA

RHODESIA
(Northern and Southern)

CA GREAT BRITAIN.
    Report to the Board of Trade on the
    trade of British South Africa (and Rhodesia) for the years .... 1911-18.

    1911: Brit. Doc. 1912-13, Cd. 6405
    1912: Brit. Doc. 1913, Cd. 7023
    1913: Brit. Doc. 1914-16, Cd. 7648
    1914: Brit. Doc. 1914-16, Cd. 8073
    1915 and 1916: Brit. Doc. 1917-18,
        Cd. 8614
    1917: Brit. Doc. 1918, Cd. 9155
    1918: Brit. Doc. 1919, Cmd. 357

RHODESIA, NORTHERN

SA RHODESIA, NORTHERN. Office of the
    Commissioner for Northern Rhodesia
    in the United Kingdom.
    Report.

    HL 1955-63   J725.3
                               A11

SA GREAT BRITAIN. Colonial Office.
    Annual report on Northern Rhodesia.
    1946-
    Previously issued in its numbered series, Colonial reports--annual (325.342 G787) no. 1292: 1924/25-no. 1935: 1938. Suspended 1940-46.

    1946-62   325.342
                               G7871Nr

    HL 1962   DT946
                               G786

SA RHODESIA, NORTHERN. Information Dept.
    The Northern Rhodesia handbook.

    HL 1950, 1953   DT963
                               A32

SA RHODESIA, NORTHERN.
    Blue book.
    Began publication with vol. 1, 1924.
    Discontinued(?)

    1929, 1931, 1938, 1948   354.689
                               R477

SA GREAT BRITAIN. Dept. of Overseas
CA   Trade.
    ... Economic conditions in East Africa and in Northern Rhodesia and Nyasaland ... Report .... [1921-38.]
    Title varies.
    Includes Uganda Protectorate, Kenya Colony and Protectorate, Tanganyika Territory, and Zanzibar Protectorate.

    1921-28, 1934/36-   380.005
      1937/48   G786

SB Central African Statistical Office.
    Economic and statistical bulletin [of] Northern Rhodesia.
    Began publication in 1948; ceased with v. 6, Mar. 1954.
    Superseded by its Monthly digest of statistics (see Rhodesia and Nyasaland).

    v. 3, no. 10-v. 6; Jan.   316.89
      1951-Mar. 1954   R475c

SA GREAT BRITAIN. Dept. of Overseas
CA   Trade.
    ... Report on economic and commercial conditions in Southern Rhodesia, Northern Rhodesia and Nyasaland ... [1932-39.]
    Statistics for Northern Rhodesia and Nyasaland, 1929/30-1930/32 are found in Economic conditions in East Africa and in Northern Rhodesia and Nyasaland ... Report, 1929/30-1930/32, respectively.

    1932, 1933, 1935, 1939   380.005
                               G786

CA Central African Statistical Office.
    Statement of the trade of Northern Rhodesia with British countries and foreign countries. 1948-49--1953.
    Previously included in the Report of the Customs Dept. of Northern Rhodesia.
    Continued in Central African Statistical Office. Statement of external trade (see Rhodesia and Nyasaland).

    1948-49--1953   382.6894
                               R477

CA RHODESIA, NORTHERN. Customs Dept.
    Report.

AR RHODESIA, NORTHERN. Ministry of African Agriculture.
    Report.
    Reports for 1960- include the reports of the Dept. of Agriculture and the Dept. of Co-operatives and African Marketing.

    1960-61, 1963   S338
                               R45A39
    FRI 1959-63

AR RHODESIA, NORTHERN. Dept. of Agriculture.
    Annual report.

    FRI 1949-58

RHODESIA, SOUTHERN

SA   RHODESIA, SOUTHERN. Dept. of Statistics.
      Statistical year book of Southern Rhodesia. 1924-   Irregular.
      Year book of 1947 includes the years 1938 to 1946.

      1947   HA1977
                                    R5A29

SA   RHODESIA, SOUTHERN. Dept. of Statistics.
      Statistical handbook of Southern Rhodesia.

      FRI   1945

SA   GREAT BRITAIN. Dept. of Overseas
CA   Trade.
      ... Report on economic and commercial conditions in Southern Rhodesia, Northern Rhodesia and Nyasaland ....
      [1932-39.]
      Title varies.

      1932, 1933, 1935, 1939   380.005
                                       G786

SA   RHODESIA, SOUTHERN.
      Official year book of the colony of Southern Rhodesia. v. 1, 1924 (no. 2 issued in 1930 for the years 1924-28).

      HL   1932   DT962
                                     A247

SB   RHODESIA, SOUTHERN. Central African Statistical Office.
      Economic and statistical bulletin of Southern Rhodesia. n.s. 1933/34-1954//
      Issued -Oct. 7, 1936, by the Government Statistical Bureau; Oct. 21, 1936- by the Dept. of Statistics; 1949-54 by the Central African Statistical Office. Ceased publication with issue for Mar. 21, 1954.

      n. s. v. 17, no. 6-v. 21,   316.89
        no. 24; June 21, 1949-   R477e
        Mar. 21, 1954

CA   RHODESIA, SOUTHERN. Customs and Excise Dept.
      Report of the controller of customs and excise.
      Contains full reports of imports and exports.

      1919-23, 1927, 1931   382.6891
                                       R475

CA   RHODESIA, SOUTHERN. Dept. of Statistics.
      Statement of the trade of Southern Rhodesia with British countries and foreign countries.
      Began publication with the 1930 issue; ceased with the issue for 1953.
      Issues for 1906-09 included in the Annual statement of the trade and shipping of the colonies and territories forming the South African Union; 1910-29 in the Annual statement of the trade and shipping of the Union of South Africa (see South Africa).
      Continued in Central African Statistical Office. Statement of external trade (see Rhodesia and Nyasaland).

      1933-34, 1946-53   382.6891
                                     R477

CA   RHODESIA, SOUTHERN. Dept. of Commerce and Industry.
      Report.

      1945-47   338.096891
                                     R477

AA   RHODESIA, SOUTHERN.
      Report on the agricultural and pastoral production of Southern Rhodesian and European farmers.

      FRI   1946/47, 1952/53

AR   RHODESIA, SOUTHERN. Ministry of Agriculture.
      Report of the Secretary for Agriculture for the year ....

      FRI   1919-22, 1962-63+

AR	RHODESIA, SOUTHERN. Dept. of Agriculture and Lands. Report. 1941/45, 1946/47-1951/52	630.96891 R477	AA	RHODESIA AND NYASALAND. Central Statistical Office. Agricultural production in Southern Rhodesia, Northern Rhodesia, and Nyasaland. FRI 1963
AR	RHODESIA, SOUTHERN. Dept. of Agriculture. Report. FRI 1919-22		AR	RHODESIA AND NYASALAND. Ministry of Agriculture. Report of the Secretary of Agriculture to the Federal Minister. FRI 1954-62

RHODESIA AND NYASALAND

SB	Central African Statistical Office. Monthly digest of statistics [of the] Federation of Rhodesia and Nyasaland. Apr. 1954-Mar. 1964// v. 1-6, Apr. 1954- Mar. 1964	316.89 C397

AR	Central African Statistical Office. Report on the agricultural [and pastoral] production of Southern Rhodesia. FRI 1946/47	

RHODESIA (1964–)

CA	Central African Statistical Office. Statement of external trade, Federation of Rhodesia and Nyasaland. 1954-63// Continues the trade report of Nyasaland, Northern Rhodesia and Southern Rhodesia, each formerly published separately. Continued in separate publications of Rhodesia, Malawi, and Zambia. 1954-63	382.689 C397
CB	RHODESIA AND NYASALAND. Central Statistical Office. External trade statistics. Quarterly. Jan./Sept. 1954-63// Continued in publications of Rhodesia and Zambia. Issues for Jan./Sept. 1954-Jan./Mar. 1959 published by the office under its earlier name: Central African Statistical Office. Jan. 1954-June 1961	382.689 R477

SB	RHODESIA. Central Statistical Office. Monthly digest of statistics. Apr. 1964- Supersedes in part the Monthly digest of statistics of the Federation of Rhodesia and Nyasaland, published by the Central African Statistical Office. Apr. 1964-Feb. 1966+	HA1977 R5A32
CA	RHODESIA. Central Statistical Office. External trade statistics. Annual. 1964- Supersedes in part Rhodesia and Nyasaland. Central Statistical Office. External trade statistics. 1964+	HF265 R5A4
CB	RHODESIA. Central Statistical Office. External trade statistics. Quarterly. Jan./Mar. 1964- Supersedes in part Rhodesia and Nyasaland. Central Statistical Office. External trade statistics (quarterly). Jan./Mar. 1964-	HF265 R5A42

AR   RHODESIA. Ministry of Agriculture.
      Report of the Secretary to the Ministry of Agriculture.

      FRI   1964

RUMANIA

SA   RUMANIA. Direcţia Centrala de Statistica.
      Anuarul statistic al R. P. R. 1904-
      Title and agency vary.
      Issued with a supplement; English translation of text.

      1909, 1915/16, 1922, 1926,   HA1641
          1931/32, 1934, 1935/36,   A5
          1939/40, 1957-65+

      HL   1912, 1915/16,   HA1641
               1922-26, 1930,   A2
               1957-65+

SA   RUMANIA. Direcţia Centrala de Statistica.
      Rumanian statistical pocket book.
      1960-

      1960, 1962-65+   314.98
                                        R936r

      HL   1960-62   HA1641
                                        A212

SA   RUMANIA. Direcţia Centrala de Statistica.
      Breviarul statistic al R. P. R. 1960-
      Issued also in English as its Rumanian statistical pocket book.

SA   RUMANIA. Institutul Central de Statistica.
      Breviarul statistic al Romaniei.
      1938-   Brief statistical summary of Rumania.

      FRI   1940

SA   GREAT BRITAIN. Dept. of Overseas
CA   Trade.
      ... Report on economic conditions in Roumania .... [1919-37.]
      1919 issued in the series of Parliamentary Papers as Parliament. Papers by command. Cmd. 828.
      Report year ends Mar. or Apr. 1919- None issued 1925, probably replaced by report dated Mar. 1926. Have two numbers for 1926; one issued with date "March 1926" and the other "during the year 1926." The latter was published in 1927; probably nothing more issued in 1927.

      1919, 1921-24, 1926,   380.005
          1928-34, 1936, 1937   G786

SB   RUMANIA. Direcţia Centrala de Statistica.
      Revista de statistica.

      Jan. 1962-Apr. 1966+   314.98
                                        R454

SB   RUMANIA. Direcţia Centrala de Statistica.
      Buletin statistic trimestrial. 1957-

      1957-no. 4, 1965+   314.98
                                        R936b

SB   RUMANIA. Institutul Central de Statistica.
      Comunicari statistice. Jan. 1945-
      Irregular.

      HL   Jan. 31, 1948

SB   RUMANIA. Institutul Central de Statistica.
      Buletin statistic al Romaniei. Bulletin statistique. 1892-

      Ser. 4, v. 14-16;   314.98
          1919-22, 1931: 1-4   R936

      HL   May/June 1938

CA   RUMANIA. Direcţiunea Statisticei
     Generale.
     Comerţul exterior. 1871-

     1910, 1912, 1914-15,   382.4981
     1919-22, 1923   R936c

CA **ROUMANIA. Direcţiunea Statisticei
     Generale a Finceior.
     Tabele grafice relative la comerţul
     exterior .... Diagrammes relatifs au
     commerce extérieur de la Roumanie.
     1871-1909, 1, 1907//

CB **ROUMANIA. Institutul de Statistica
     Generala a Statulul.
     Buletinul semestrial al statisticei
     comerciale .... Bulletin. no. 1-2,
     1919//

CB **ROUMANIA. Direcţiunea Vamilor si
     Statisticei Financiare.
     Buletin statistic al comerciuiul
     Romaniei cu terile streine .... Bulletin
     statistique du commerce ... avec les
     pays étrangers .... Jan. 1891-Dec.
     1893//

AA   RUMANIA. Ministerul Agriculturei si
     Domeniilor. Direcţiunea Statisticei
     Generale.
     Statistica agricola.

     FRI 1911-15, 1918, 1920:
          pt. 1, 1923/24: pt. 1-
          1938, 1941/44-1947

AA   ROUMANIA. Ministerul Agriculturei si
     Domeniilor.
     L'agriculture en Roumanie. Album
     statistique 1929//

     1929   630.9498
                                   R936

AA   RUMANIA. Ministerul Agriculturei si
     Domeniilor.
     Statistica semanaturilor din regatul
     roman in anul agricol ....

     HL   1919-20

AA **ROUMANIA. Institutul de Statistica
     Generala a Statulul.
     Statistica agricola a Romaniei ...
     Exploatarile agricole 1911-15, 1,
     1918//

AA **ROUMANIA. Ministerul Agriculturei si
     Domeniilor.
     Anuarul 1859/93, 1900/01,
     1911/12//

AA   RUMANIA. Ministerul Agriculturei,
     Industriei, Comerciului si Domeniilor.
     La Roumanie, 1866-1906.

     1866-1906 (1 vol.)   330.94981
                                   R936

AB   RUMANIA. Ministerul Agriculturei si
     Domeniilor.
     Buletinul informativ.

     Anul 3: 1-11: 1,   630.6498
     Jan. 1932-Jan. 1940   R935

AB   RUMANIA. Ministerul Agriculturei si
     Domeniilor.
     Buletinul agriculturei. 1920-
     Supersedes its Buletinul, 1885-1916.
     Some vols. accompanied by Supple-
     ments.

     FRI 1920: no. 4-1930
          wanting: several issues

RUSSIA

SA   RUSSIA (1917-   R.S.F.S.R.). TSen-
     tral'noe Statisticheskoe Upravlenie.
     RSFSR: kratkii statisticheskii spra-
     vochnik.

     HL   1959, 1962-64+   HA1432
                                    A66

SA   RUSSIA (1923-   U.S.S.R.). TSen-
      tral'noe Statisticheskoe Upravlenie.
      SSSR v tsifrakh; kratkii statisticheskii
      sbornik. 1959-
      Vol. for 1959 includes material for
      1913-59.

      HL   1959-64+   HA1432
                                A67

      FRI   1959-60   330.947
                                R9675st

SA   RUSSIA (1923-   U.S.S.R.). TSen-
      tral'noe Statisticheskoe Upravlenie.
      Narodnoe khoziaistvo SSSR; statisti-
      cheskii ezhegodnik.

      1958, 1960-61, 1963-64+   330.947
                                N233

      HL   1956, 1958-64+   HC336.4
                                N233

SA   RUSSIA (1923-   U.S.S.R.). TSen-
      tral'noe Statisticheskoe Upravlenie.
      U.S.S.R. in figures for 1959; brief
      statistical returns.
      "Data from the period 1913-1959."

      1959   314.7
                                R972a

      HL   1959   HA1432
                                A67

SA   RUSSIA (1917-   R.S.F.S.R.). TSen-
      tral'noe Statisticheskoe Upravlenie.
      RSFSR za sorok let. 1957.

      HL   1957   HA1435
                                A211

SA   RUSSIA (1923-   U.S.S.R.). TSen-
      tral'noe Statisticheskoe Upravlenie.
      Narodnoe khoziaistvo SSSR; statisti-
      cheskii sbornik.

      1956   330.947
                                R9675

      HL   1956   HC336.4
                                A33

SA   RUSSIA (1923-   U.S.S.R.). TSen-
      tral'noe Statisticheskoe Upravlenie.
      National economy of the USSR; statis-
      tical yearbook.
      English edition of its Narodnoe
      khoziaistvo SSSR; statisticheskii ezhe-
      godnik.

      HL   1956   HC336.4
                                A332b

SA   RUSSIA (1923-   U.S.S.R.). TSen-
      tral'noe Statisticheskoe Upravlenie.
      The national economy of the USSR; a
      statistical compilation.

      1956   Microfilm
                                330
                                2

SA   RUSSIA (1923-   U.S.S.R.). TSen-
      tral'noe Statisticheskoe Upravlenie.
      Supplement to "The national economy
      of the USSR; a statistical compilation."
      "The data have been extracted from
      the periodical Vestnik statistiki, no. 5,
      September/October 1956, pages 59-95."

      1956   330.947
                                R9675bs

SA   RUSSIA (1923-   U.S.S.R.). TSen-
      tral'noe Statisticheskoe Upravlenie.
      SSSR v tsifrakh. 1913-57 (1 vol.).

      HL   1913-57   HA1432
                                A5
                                1958

SA   RUSSIA (1923-   U.S.S.R.). TSen-
      tral'noe Upravlenie Narodno-
      Khoziaistvennogo Ucheta.
      Kratkii statisticheskii spravochnik.
      1936.

      HL   1936   HA1435
                                A5
                                1936

SA RUSSIA (1923- U.S.S.R.). TSen-
     tral'noe Upravlenie Narodno-
     Khoziaistvennogo Ucheta.
     Sotsialisticheskoe stroitelstvo Soiuza
     SSR 1933-1938gg; statisticheskii sbor-
     nik.
         Socialist construction in the U.S.S.R.
     1933-1938; complete statistical abstract.

     HL   1934-36 (3 vol.,   HA1435
          microfilm)   A517

SA RUSSIA (1923- U.S.S.R.). TSen-
     tral'noe Upravlenie Narodno-
     Khoziaistvennogo Ucheta.
     The U.S.S.R. in figures. 1934, 1935.
         1934, "A summary of the book 'Social-
     ist construction in the U.S.S.R.'"
         1935, new material added to that con-
     tained in the 1934 vol.

     HL   1934-35   HA1432
                                  A52

SA RUSSIA (1923- U.S.S.R.). TSen-
     tral'noe Statisticheskoe Upravlenie.
     Statisticheskii spravochnik SSSR.
     1924-
         Vol. for 1926 never published.

     HL   1924-25, 1928   HA1433
                                  A75

SA RUSSIA (1923- U.S.S.R.). TSen-
     tral'noe Statisticheskoe Upravlenie.
     Ten years of Soviet power in figures,
     1917-1927.
         "... replaces the statistical year
     books of the U.S.S.R. for the years
     1925-1927, which were not issued."

     HL   1917-27   HA1439
                                  A6
                                  1927

     FRI 1917-27   HA1432
                                  A5
                                  1927

SA RUSSIA (1923- U.S.S.R.). TSen-
     tral'noe Statisticheskoe Upravlenie.
     Abrégé des données statistiques de
     l'Union des républiques socialistes
     soviétiques, rédigé pour les membres de
     la XVI session de l'Institut international
     de statistique par l'Administration cen-
     trale de statistique de l'U.R.S.S. 1925.

     HL   1925   HA1435
                                  R969

     FRI 1925   HA1435
                                  A5

SA RUSSIA (1923- U.S.S.R.). TSen-
     tral'noe Statisticheskoe Upravlenie.
     Statisticheskii ezhegodnik. 1913-24.
     Subseries of its Trudy.
         1913-17 have title: Statisticheskii
     sbornik.

     HL   1913-24   HA1432
                                  A2
                                  v. 7-8

SA RUSSIA. TSentral'nyi Statisticheskii
     Komitet.
     Statisticheskii ezhegodnik Rossii.
     1904-16//
         Title varies: 1904-10, Ezhegodnik
     Rossii.
         French title on verso of title page:
     1904-10, Annuaire de la Russie; 1911-16,
     Annuaire statistique de la Russie.

     HL   1904, 1908-16   HA1432
                                  A1

SB RUSSIA (1923- U.S.S.R.).
     Vestnik statistiki; organ TSentral'nogo
     statisticheskogo upravleniia pri Sovete
     ministrov SSSR.

     HL   no. 1, Jan./Feb. 1949   Microfilm
                                  S51

          no. 4-6, 1955; no. 1,   Russian
          1956-no. 12, 1965+   Serials

CA RUSSIA (1923- U.S.S.R.). Planovo-
     ekonomicheskoe Upravlenie.
     Vneshniaia torgovlia Soiuza SSR;
     statisticheskii obzor.
         "Prilozhenie k zhurnalu 'Vneshniaia
     torgovlia.'"

     FRI 1958-59   HF207
                                  V6

CA RUSSIA (1917- R.S.F.S.R.). TSentral'noe Statisticheskoe Upravlenie.
    Sovetskaia torgovlia v RSFSR. 1958.

    HL   1958   HF3621
                             A35

CA RUSSIA (1923- U.S.S.R.). TSentral'noe Statisticheskoe Upravlenie. Otdel Statistiki Torgovli.
    Sovetskaia torgovlia. 1956.

    HL   1956   HF3621
                             A47

CA RUSSIA (1923- U.S.S.R.). Glavnoe Tamozhennoe Upravlenie.
    Vneshniaia torgovlia SSSR za 1918-1940gg; statisticheskii obzor.

    FRI 1918-40 (1 vol.)   HF207
                             V59

AA RUSSIA (1923- U.S.S.R.).
    [Agriculture of the Soviet Union for the years 1953-1962].
    Issued in its Vestnik statistiki; organ TSentral'nogo statisticheskogo upravleniia pri Sovete ministrov SSR, v. 9: 79-95, 1963.

    HL   v. 9: 79-95, 1963

AA RUSSIA (1923- U.S.S.R.). TSentral'noe Statisticheskoe Upravlenie.
    Aperçu statistique sur l'agriculture en URSS pour la période ....

AA RUSSIA. Direction Générale de l'Organisation Agraire et de l'Agriculture. Division d'Economie Rurale et de Statistique Agricole.
    Recueil de données statistiques et économiques sur l'industrie agricole en Russie et dans les pays étrangers.

    FRI 1918

AA RUSSIA. Ministère de l'Agriculture. Division d'Economie Rurale et de Statistique Agricole.
    Recueil de données statistiques et économiques sur l'industrie agricole en Russie et dans les pays étrangers.

    FRI 1913, 1915

AA RUSSIA. TSentral'nyi Statisticheskii Komitet.
    Résultats généraux de la récolte en Russie. 1883/87-
    Title varies: 1883/87, La récolte moyenne dans la Russie d'Europe.

    1905, 1907-08   630.647
                             R967

RWANDA

SB RWANDA. Direction de la Statistique Générale et de la Documentation.
    Bulletin de statistique. 1, Apr. 1964-

    1-6, Apr. 1964-July 1965+   HA2117
                                 R9A4

CB RWANDA.
    Bulletin international des douanes.

S

SABAH

SA SABAH.
    Annual bulletin of statistics; Sabah.

    FRI 1964

SA GREAT BRITAIN. Colonial Office.
    Sabah, report.
    Annual. Published only for 1963.
    Filed and listed with reports for North Borneo.
    1964- see Annual report--Malaysia.

CB   SABAH. Dept. of Statistics.
     Sabah statistics of external trade.
     1954-

AR   SABAH. Dept. of Agriculture.
     Report. Annual.

ST. CHRISTOPHER AND NEVIS

SA   **ST. CHRISTOPHER AND NEVIS.
     Blue book. 1821-87//

AR   **ST. CHRISTOPHER AND NEVIS. Agri-
     cultural Dept.
     Report. 1914/15.
     Issued by Imperial Dept. of Agricul-
     ture for the West Indies.

ST. HELENA

SA   GREAT BRITAIN. Colonial Office.
     Annual report on St. Helena. 1947-
     Previously issued in its numbered
     series, Colonial reports--annual
     (325.342 G787) no. 1084: 1920-
     no. 1909: 1938. Suspended 1940-46.

     1947--1962-63+   325.342
                               G7871Sh

SA   ST. HELENA.
     Blue book. 1836-

ST. KITTS-NEVIS-ANGUILLA

SA   GREAT BRITAIN. Colonial Office.
     St. Kitts-Nevis-Anguilla; report.
     1955/56-
     Continues in part its Report on the
     Leeward Islands, issued in the series
     Colonial annual reports (325.342
     G7871Le).

     1955-58+   325.342
                               G7871sk

ST. LUCIA

SA   GREAT BRITAIN. Colonial Office.
     Annual report on St. Lucia, B.W.I.
     1946-
     Previously issued in its numbered
     series, Colonial reports--annual
     (325.342 G787) no. 1141: 1921-
     no. 1929: 1938. Suspended 1940-46.

     1946-62+   325.342
                               G7871SL

SA   ST. LUCIA. Statistical Unit.
     Statistical digest. Annual. 1959(?)-

SA   ST. LUCIA.
     Blue book. 1821-

     1916/17-1922, 1923-38   354.7298
                               S245b

SB   ST. LUCIA. Statistical Unit.
     Quarterly digest of statistics. 1,
     Nov. 1959-

CA   ST. LUCIA. Statistical Unit.
     Overseas trade of St. Lucia.

     1960-61+   HF157
                               B8S27

CA   ST. LUCIA. Customs, Excise and Supply
     Dept.
     Report of the customs revenue, imports
     and exports, balance of trade and excise.

     1951-59+   HF157
                               B8S25

SB   ST. LUCIA. Statistical Unit.
     Overseas trade. Monthly report. 1,
     Jan. 1960-

AA   WEST INDIES. Federal Statistical Office.
     Agricultural statistics, series 1.
     (1956-58 survey)

     FRI no. 6. The survey in
              St. Lucia. 1959

AR **ST. LUCIA. Agricultural Dept.
    Report. 1911/12-
    Continues Botanical Garden. Report.
    At head of title: Imperial Dept. of
    Agriculture for the West Indies.

AR **ST. LUCIA. Botanical Garden.
    Report on the botanic station, agricultural school and experiment plots. 1887-1910/11//
    Continued as Agricultural Dept.
    Report.

ST. PIERRE AND MIQUELON

SA **ST. PIERRE AND MIQUELON.
    Annuaire des Iles Saint Pierre et Miquelon. 1874-1914//

CA FRANCE. Institut National de la Statistique et des Etudes Economiques.
    Statistiques du commerce extérieur de Saint Pierre et Miquelon.

    1964   HF157
                                    S23F7

ST. THOMAS AND PRINCIPE

SA **ST. THOMAS AND PRINCIPE.
    Anuário comercial, industrial e agrícola da Província de S. Thomé e Príncipe. 1928-

SA ST. THOMAS AND PRINCIPE.
    See Portugal. Instituto Nacional de Estatística. Anuário estatístico; and its Anuário estatístico do império colonial (now Anuário estatístico do Ultramar).

ST. VINCENT

SA GREAT BRITAIN. Colonial Office.
    Annual report on St. Vincent. 1946-
    Previously issued in its numbered series, Colonial reports--annual

    (325.342 G787) no. 1094: 1920-no. 1933: 1938. Suspended 1940-46.

    1946--1962-63+   325.342
                                    G7871Sv

SA   ST. VINCENT.
     Blue book. 1871-

     1914/15-1937   354.7298
                                    S155b

SB   ST. VINCENT. Statistical Unit.
     Quarterly digest of statistics. 1, Nov. 1959-

CA **ST. VINCENT. Customs Office.
    Report of the supervisor of customs upon the import and export returns.
    Title varies.

CB   ST. VINCENT. Statistical Unit.
     Digest of statistics on overseas trade; finance and banking; transport....

AA   WEST INDIES. Federal Statistical Office.
     Agricultural statistics, series 1.
     (1956-58 survey)

     FRI no. 5. The survey in
          St. Vincent. 1959

AR **ST. VINCENT. Agricultural Dept.
    Report. 1911/12-
    Issued by Imperial Dept. of Agriculture for the West Indies.
    Continues Botanic Station. Reports.

AR **ST. VINCENT. Botanic Station.
    Reports on the botanic station, agricultural school, and land settlement scheme, St. Vincent. 1901/02-1910/11//
    Issued by the Imperial Dept. of Agriculture for the West Indies.
    Continued as Agricultural Dept.
    Report.

SALVADOR

SA EL SALVADOR. Dirección General de Estadística y Censos.
    Anuario estadístico. 1911-
    Some volumes issued in parts. No volumes issued 1924-26.

1911-14, 1916-20, 1922, 1927, 1930-40: 2, 1942: 2, 1943: 1, 1944: 1-3, 1948: 1-3, 1950/51-1964: 1+	317.284 S182

SA SALVADOR. Dirección General de Estadística.
    ... Resumen estadístico de la república de El Salvador. 1924-[1925].
    Issued as one of the publications which were substituted for the Anuario estadístico during 1924-26.

1924-[1925]	317.284 S1824

SB EL SALVADOR. Dirección General de Estadística y Censos.
    Boletín estadístico. 1935-
    Frequency varies.
    Publication suspended after issue for 1st quarter 1935 (no. 7) until May 1937 (no. 8) during which time all statistical information was included in Boletín of the Ministerio de Hacienda y Crédito Público.

2d ser. no. 3-no. 67, May/June 1952- Sept. 1965+ wanting: Dec. 1952	317.284 S182b

CA SALVADOR. Dirección General de Estadística y Censos.
    Comercio exterior.
    Issued through 1923 by Dirección General de Estadística (Ministerio de Hacienda y Crédito Público) with title: Estadística de comercio exterior.

1923, 1952-54, 1958$^{1-2}$, 1959^1, 1961^1	382.7284 S18

CA SALVADOR. Dirección General de Estadística y Censos.
    Estadística comercial de 1924-1925.
    Published with its Demografia, 1924-26, Division administrativa ..., 1924, and Resumen estadística ..., in place of the Anuario estadístico which was not published 1924-26.

1924-25	317.284 S1823

CA **SALVADOR. Ministerio de Hacienda.
    Sinopsis estadística de finanzas y comercio. 1919//
    Text in Spanish and English.

CA SALVADOR. Ministerio de Hacienda y Crédito Público.
    Estadística de importación y de exportación ... con forme a la nomenclatura de Bruselas.

1916-19	382.7284 S182

AB **SALVADOR. Dirección General de Agricultura.
    Boletín. 1, 1924-

AB **SALVADOR. Dirección General de Agricultura.
    Revista de agricultura tropical. 1, 1921-
    Supersedes its Boletín de agricultura y fomento.
    Suspended 1926-27.

AB **SALVADOR. Dirección General de Agricultura.
    Boletín de agricultura y fomento. v. 1, no. 1-3, Jan.-Mar. 1920//
    Superseded by its Revista de agricultura tropical.

AR SALVADOR. Ministerio de Agricultura y Ganaderia.
    Informe.
    Title varies: 1952/53-1954/55, Memoria.

1963/64-1964/65+	S174 A3

SARAWAK

SA   GREAT BRITAIN. Colonial Office.
      Annual report on Sarawak. 1947-
      Previously issued in its numbered series, Colonial reports--annual (325.342 G787). Suspended 1940-46.

      1947-62+   325.342
                          G7871Sa

SA   **SARAWAK. Treasury Dept.
CA   Annual report. 1900-
      1900-04 as Report and statement of the Treasury Department; 1905-24, Annual report of the treasury and post, shipping and customs departments (title varies slightly).

SA   SARAWAK.
      Annual bulletin of statistics, Sarawak.

CA   SARAWAK. Dept. of Statistics.
      Statistics of external trade.

      First half year, 1963-65+   HF259
      Annual, 1963-64+   S3A4

CA   SARAWAK. Dept. of Trade and Customs.
      Report. 1925-
      Issued 1964- by the Sarawak Region of Royal Customs and Excise, Malaysia.

      1964+   HJ7240
                              S3A4

AA   **SARAWAK. Dept. of Agriculture.
      Annual report.

AA   SARAWAK. Dept. of Agriculture.
      A digest of agricultural statistics.

SAUDI ARABIA

SA   SAUDI ARABIA. Central Dept. of Statistics.
      Statistical yearbook. 1965-

      1965+   HA1950
                              S3A4

      HL   1965+   HA1687
                              S2A25

CA   SAUDI ARABIA. Central Dept. of Statistics.
      Summary of foreign trade statistics.
      English and Arabic.

      1963/64-1964/65+   HF259
                                S35A4

AA   SAUDI ARABIA. Ministry of Agriculture.
      Ministry of Agriculture during one year.
      Issued by the ministry's Division of Statistics and Agricultural Economics. English and Arabic.

      1963/64+   S322
                              S35A4

AR   SAUDI ARABIA. Ministry of Agriculture.
      Report.
      English and Arabic.

      1963/64+   S322
                              S35A3

SCOTLAND

see also statistics for United Kingdom published and listed under GREAT BRITAIN

SB   SCOTLAND. Scottish Statistical Office.
      Digest of Scottish statistics.
      Began publication with the Apr. 1953 issue.

      no. 13-26, Apr. 1959-   314.1
         Oct. 1965+   S424

AA   SCOTLAND. Dept. of Agriculture for
        Scotland.
        Agriculture in Scotland.
        Published in series: Gt. Brit. Par-
   liament. Papers by command.

      1959-64+   Brit. Docs.

AA   SCOTLAND. Dept. of Agriculture and
        Fisheries.
        Agricultural statistics. 1912-
        Issued through 1927 by the Board of
   Agriculture.
        Vols. for 1912-20 issued as Papers
   by command.

      1912-20   Brit. Docs.
      1921, 1923, 1933, 1936-38,   630.641
        1939/44: pt. 1, 1945-64+   S424ag

AA   SCOTLAND. Dept. of Agriculture.
        Abstract of agricultural returns for
   Scotland. 1912-
        Issued through 1927 by the Board of
   Agriculture.

      1913-15, 1918-34   630.641
                                    S424ar

AR   SCOTLAND. Dept. of Agriculture.
        Report. 1912-
        Issued as Papers by command.

      1912-64+   Brit. Docs.

AR **SCOTLAND. Dept. of Agriculture.
        Monthly agricultural report. 1912-
        Title varies.

SENEGAL

see also FRENCH WEST AFRICA

SA   SENEGAL. Service de la Statistique.
        Situation économique du Sénégal.

      1962-64+   HC547
                                    S4A55

SA **SENEGAL.
        Annuaire du Sénégal et dépendances.
   1854-1902//
        Continued as French West Africa.
   Annuaire du gouvernement general de
   l'Afrique occidentale française.

SB   SENEGAL. Service de la Statistique et
        de la Mécanographie.
        Bulletin statistique et économique
   mensuel. 1959-
        Issued in 1959 by the agency under a
   variant name: Service de la Statistique
   Générale.
        Supersedes in part Bulletin statistique
   issued by French West Africa. Service
   de la Statistique du Sénégal et de la
   Mauritanie, 1953-58.
        Title varies: 1959, Bulletin statis-
   tique; Jan./Feb. 1960, Bulletin bimes-
   triel statistique et économique.

      3-4, 5, 6, 1959;   HA2117
      1, Jan./Feb. 1960;   S4A3
      1, Sept./Oct. 1960;
      1961-Sept. 1965--

      HL   3-4, 1964

CB   SENEGAL. Service de la Statistique et
        de la Mécanographie.
        Commerce extérieur du Senegal:
   commerce spécial.
        Monthly with each issue cumulative
   through the year.

      Jan. 1962-Jan. 1966+   382.663
                                          S475

SEYCHELLES

SA   GREAT BRITAIN. Colonial Office.
        Annual report on Seychelles. 1946-
        Previously issued in its numbered
   series, Colonial reports--annual
   (325.342   G787) no. 1106: 1920-no. 1890:
   1937. Suspended 1940-46.

      1946-1963/64+   325.342
                                          G7871Sy

SA   SEYCHELLES.
        Blue book. 1899-1939(?)

CA   SEYCHELLES.
      Trade report.

     1961-64+   HF266
                                 S4A3
                                 f

AB **SEYCHELLES. Dept. of Agriculture and
      Fisheries.
      Bulletin. 1, 1923.
      Articles in French and English.

AR   SEYCHELLES. Dept. of Agriculture.
      Annual report. 1913-
      1913-21 as Botanic Station. Annual
report on agriculture and crown lands.

     FRI   1959, 1962-64+

SIERRA LEONE

SA   GREAT BRITAIN. Colonial Office.
      Annual report on Sierra Leone. 1946-
     Previously issued in its numbered
series, Colonial reports--annual
(325.342 G787) no. 1090: 1920-
no. 1916: 1938. Suspended 1940-46.

     1946-58+   325.342
                                 G7871Si

SA   GREAT BRITAIN. Dept. of Overseas
CA   Trade.
      ... Report on economic and commer-
cial conditions in West Africa (the Gambia, Sierra Leone, the Gold Coast and
Nigeria) .... 1936/37.

     1936/37   380.005
                                 G786

SA   SIERRA LEONE.
      Blue book. 1819-1938(?)

     1913-28   354.664
                                 S572

SB   SIERRA LEONE. Central Statistics
      Office.
      Quarterly statistical bulletin. no. 1,
Mar. 1963-

     no. 1-4, 1963-Dec. 1964+   HA1977
                                          S5A3
                                          f

CA   SIERRA LEONE. Customs Dept.
      Trade report.

     1926, 1945-52, 1954-63+   382.664
                                          S572

CB   SIERRA LEONE.
      Quarterly trade statistics.

     no. 425-436, Jan./Mar.   HF266
       1962-Oct./Dec. 1964+   S5A3

AR   SIERRA LEONE. Agricultural Dept.
      Annual report.

     FRI   1929, 1952-53,
               1955-59

AR   SIERRA LEONE. Lands and Forests
      Dept.
      Annual report.
      Jan. 1, 1929, the Lands and Forests
Dept. was split into two departments, the
Agricultural Dept. and the Forestry Dept.

     1923-28   634.9664
                                 S572

AR **SIERRA LEONE. Dept. of Agriculture.
      Annual report. 1912-18//
      Continued in Lands and Forest Dept.

SINGAPORE

SA   SINGAPORE.
    Report. 1946-
    Issued also in the series, Colonial annual reports (Gt. Brit. Colonial Office).
    Previously issued in the numbered series, Colonial reports--annual, which was suspended 1940.

1946-60+   325.342
                               G7871Sn

SA   SINGAPORE. Dept. of Statistics.
    Blue book. 1946-

1946   315.95
                                S617

SB   SINGAPORE. Dept. of Statistics.
    Digest of statistics. 1962-
    Supersedes in part Malayan statistics; digest of economic and social statistics (State of Singapore and Federation of Malaya).

v. 1, no. 1, Jan. 1962-   315.95
  Oct. 1965+   S617d
wanting: June 1963

SB   SINGAPORE. Dept. of Statistics.
    Malayan statistics; digest of economic and social statistics (State of Singapore and Federation of Malaya).
    Superseded by its Digest of statistics.

Mar./Sept. 1947, 1948,   315.95
  Jan./June 1949, Apr.   M239ss
  1950-Dec. 1961
wanting: Feb. 1954

CA   SINGAPORE. Dept. of Statistics.
    External trade. 1956-

CB   SINGAPORE. Dept. of Statistics.
    External trade statistics (excluding trade between Singapore and the Federation of Malaya).

(continued)

Began publication with the July 1921 issue; suspended Nov. 1941-July 1946. Title varies.

1921, 1923, 1928, Mar.   382.595
  1947-1965: no. 2+   M241
wanting: Oct./Dec. 1947,
  July/Dec. 1949

CA   SINGAPORE. Dept. of Statistics.
    Summaries of Malayan foreign trade (supplement to Malayan statistics).

Oct. 1946-Mar. 1950   382.595
                                 S617

SOMALIA

SA   SOMALIA. Planning Directorate. Dept. of Statistics.
    Statistical abstract for Somalia. no. 1, 1964-

CA   SOMALIA. Northern Region. Customs and Excise Dept.
    Annual trade report. 1962-

CA   SOMALIA. Servizio di Statistica.
    Statistica del commercio con l'estero. 1960-

SOMALILAND, BRITISH

SA   GREAT BRITAIN. Colonial Office.
    Annual report on the Somaliland Protectorate.
    Previously issued in its numbered series, Colonial reports--annual (325.342 G787) no. 1100: 1920- no. 1880: 1937. Suspended 1940-46.

1948-59   325.342
                               G7871Som

SA   **SOMALILAND PROTECTORATE.
    Blue book. 1910/11-

CA   SOMALILAND, BRITISH. Customs and
       Excise Dept.
       Trade report of the Somaliland Pro-
tectorate. 1951-

CA **GREAT BRITAIN. Colonial Office.
       Report on the trade of the Somaliland
Protectorate. 1891/92-
       To 1904/05 issued by the Foreign
Office in the series: Diplomatic and con-
sular reports.

SOMALILAND, FRENCH

SA   FRANCE.
       Côte française des Somalis. 1954.

     1954   HA2117
                                   S6F7

SOMALILAND, ITALIAN

SA **SOMALILAND, ITALIAN.
       Bollettino ufficiale della Somalia
italiana. 1, 1910-

SB   SOMALILAND, ITALIAN.
       Bollettino ufficiale dell'administra-
zione fiduciaria italiana della Somalia.
v. 1-11, Apr. 1950-60.
       Supplements accompany most numbers.

SB   SOMALIALAND, ITALIAN.
       Bollettino ufficiale. v. 1-4, no. 6,
Jan. 1957-June 1960.
       Supplements accompany some num-
bers.
       Superseded by Bollettino ufficiale
issued by Somalia.

SB   SOMALILAND, ITALIAN.
       Bollettino mensile. Mar. 1951-

CA   ITALY. Ministero dell'Africa Italiana.
       Ufficio Studi e Propaganda.
       ... Statistica del movimento commer-
ciale marittimo dell'Eritrea, della
Somalia italiana, della Tripolitania e
della Cirenaica e del movimento commerciale carovaniero dell'Eritrea ....
1921 e 1922-
       1925-26 includes "Movimento com-
merciale marittimo dell'Oltre Giuba dal
1° luglio 1925 al 20 giugno 1926."
       1928-29 includes "Movimento della
navigazione marittima della quattro
colonie."

     1923-24   382.45
                                   I892

HL   1923-24, 1931-32   HF270
                                   I88

CA   SOMALILAND, ITALIAN. Ufficio Central
       di Statistica.
       Statistica del commercio con l'estero.

SOUTH AFRICA

SA   SOUTH AFRICA. Bureau of Statistics.
       Statistical year book. Statistiese
jaarboek. 1964-
       Supersedes in part the Official year
book of the Union and of Basutoland,
Bechuanaland Protectorate and Swaziland.

     1964+   HA1991
                                   A5

HL   1964+   HA1991
                                   A3

SA   SOUTH AFRICA. Bureau of Census and
       Statistics.
       Official year book of the Union and of
Basutoland, Bechuanaland Protectorate
and Swaziland. 1917-60//
       Title varies.
       Supersedes its Statistical year book.
       Superseded in part by Statistical year
book issued 1964 by the Bureau of Statistics.
       Vol. 1, statistics for 1910-16.

     v. 1-30, 1917-60   316.8
                                   S725

SA   SOUTH AFRICA. Bureau of Census and
          Statistics.
       Uniestatistieke oor vyftig jaar. Union
    statistics for fifty years. 1910-60.

       1910-60 (1 vol.)   316.8
                                   S725u

       HL 1910-60 (1 vol.)   HA1992
                                   A5

SA   SOUTH AFRICA.
       ... Report of the government of the
    Union of South Africa on South-West
    Africa for the year ....

       HL 1918-38, 1946   DT714
           wanting: 1930   A5S728

SA   SOUTH AFRICA. Bureau of Census and
          Statistics.
       Statistical year book of the Union of
    South Africa. no. 1-4, 1912/13-1915/16//
       Superseded by its Official year book of
    the Union and of Basutoland, Bechuana-
    land Protectorate and Swaziland, 1917-

       no. 1-4, 1912/13-1915/16   316.8
                                   S725s

SB   SOUTH AFRICA. Bureau of Census and
          Statistics.
       Monthly bulletin of statistics.

       Dec. 1928 (no. 85)-   316.8
         Mar. 1938, Jan. 1961-   S725m
         Jan. 1962, Mar. 1965-
         Mar. 1966+

SB   SOUTH AFRICA. Bureau of Census and
          Statistics.
       Quarterly abstract of Union statistics.
    no. 1-13, Jan. 1920-Jan. 1923//
       Supersedes its Half-yearly abstract.

       no. 1-13, Jan. 1920-23   316.8
                                   S725q

SB   SOUTH AFRICA. Bureau of Census and
          Statistics.
       Half-yearly abstract of Union statis-
    tics .... no. 1, June 1919//
       Superseded by its Quarterly abstract.

       no. 1 and suppl.   316.8
                                   S725h

CA   SOUTH AFRICA. Dept. of Customs and
          Excise.
       Annual statement of the trade and
    shipping of the Union of South Africa and
    the territory of South-West Africa.
       Began publication in 1910, in continua-
    tion of the Annual statement issued for
    1906-09 by the South African Customs
    Statistical Bureau, Cape Town.
       Customs tariff included, 1910-26.
       Title varies.

       1910-23, 1945-64: v. 1+   382.68
                                   S725

CB   SOUTH AFRICA. Dept. of Customs and
          Excise.
       Monthly abstract of trade statistics.

       1947-June 1965+   382.68
                                   S725m

AA   SOUTH AFRICA. Bureau of Census and
          Statistics.
       Report on agricultural and pastoral
    production.
       Began publication in 1918; suspended
    1940-45.
       Agency name and title vary.

       FRI 1918-22, 1922/23-
              1929/30, 1933/34,
              1936/37, 1945/46,
              1946/47, 1954/55-
              1957/58

AR   SOUTH AFRICA. Dept. of Agriculture.
       Report with appendices.

       1910/11-1919/20, 1921/23   630.668
       wanting: 1914/15-1915/16   S725

       FRI 1945-47

SOUTH ARABIA

see FEDERATION OF SOUTH ARABIA

SPAIN

SA   SPAIN. Instituto Nacional de Estadística.
      Anuario estadístico de España. 1912-
      Publication suspended 1935-42.
      Vols. for 1944-45 and 1946-47 are combined issues.
      Issuing agency varies.

      1915-20, 1923-24, 1929-30,   HA1543
        1944/45, 1947, 1950,   A5
        1953, 1959, 1963-65+

      HL   1943/44, 1946/47,   HA1543
            1948   A3

SA   SPAIN. Instituto Nacional de Estadística.
      Anuario estadístico de España; edición manual.

      1941, 1942-53, 1959-63+   HA1543
                                A52

      HL   1941, 1945/46, 1947,   HA1541
            1948/49   A636

SB   SPAIN. Instituto Nacional de Estadística.
      Estadística española; revista del Instituto Nacional de Estadística. no. 1, Oct./Dec. 1958-

      Apr./June 1959-   HA1
        Oct./Dec. 1965+   E82

SB   SPAIN. Instituto Nacional de Estadística.
      Boletín de estadística.
      Began publication with the Jan./Mar. 1939 issue.
      Issued 1939-45 by the Dirección (called Servicio, Jan./Mar.-Apr./June 1959) General de Estadística.
      No. 37- called "2. época, año 10- "

      no. 17-228, Mar. 1943-   314.6
        Dec. 1963+; suppl. 1-9,   S733b
        1947-57
      wanting: no. 23, 25-28, 31

CA   SPAIN. Dirección General de Aduanas.
      Estadística del comercio exterior de España.
      Title varies: -1855, Cuadro general del comercio esterior de España;   --1917-19, Estadística general del comercio exterior de España.

      1917-19, Jan.-June 1944,   382.46
        Jan.-June 1945, 1948,   S733
        1951, 1953: v. 2,
        1963: v. 1-2+

CA **SPAIN. Consejo de la Economía Nacional.
      Resumen estadístico del comercio exterior de España. 1, 1924-
      Continues Dirección General de Aduanas. Resúmenes mensuales de la estadística del comercio exterior.

CB   SPAIN. Dirección General de Aduanas.
      Estadística del comercio exterior de España. Trimestral.

      1st trim. 1965+

CB   SPAIN. Dirección General de Aduanas.
      Estadística del comercio exterior de España; resumen mensual.

      Nov. 1950-Feb. 1954

CB **SPAIN. Dirección General de Aduanas.
      Resúmenes mensuales de la Estadística del comercio exterior de España....
      no. 1-408, 1899-1923//
      Continued as Consejo de la Economía Nacional. Resumen estadístico del comercio exterior.

AA   SPAIN. Ministerio de Agricultura. Servicio de Estadística.
      La agricultura española; información estadística y económica.

      1964+   S253
                              A52

      FRI   1963-65+

AA   SPAIN. Ministerio de Agricultura. Servicio de Estadística.
    Anuario estadístico de las producciónes agrícolas.

    FRI 1933-35, 1943-July
            1946, 1948-
            1963/64+

AA **SPAIN. Consejo Agronómico.
    Anuario estadístico-agrícola. no. 1-3, 1928//
    Continued as Dirección General de Agricultura .... Anuario estadística de las producciónes agrícolas.

AB   SPAIN. Ministerio de Agricultura.
    Información estadística y económica.
Monthly.

    FRI Dec. 1962-Dec. 1965+

AB **SPAIN. Dirección General de Agricultura, Industria y Comercio.
    Boletín quincenal de estadística, mercados e informaciónes agrícolas. 1-5, (no. 1-116), 1903-Dec. 25, 1907//
    Continues Boletín semanal.

AB **SPAIN. Dirección General de Agricultura, Industria y Comercio.
    Boletín semanal de estadística y mercados. 1-12, 1892-1902//
    Continued as Boletín quincenal.

SPAIN, COLONIES

SA   SPAIN. Direccion General de Plazas y Provincias Africanas.
    Resumen estadístico del Africa española.

    1954   312.6
                         S733
    HL 1954, 1956/58,
          1961/62

SPANISH GUINEA

see GUINEA, SPANISH

SPANISH MOROCCO

SA   SPAIN. Instituto Nacional de Estadística.
    Zona de protectorado y de los territorios de soberanía de España en el norte de Africa. Anuario estadístico. 1941-
    Issued 1941-45 by the Dirección General de Estadística.

    1944, 1949-50, 1951,   316.4
      1952-53   S733

SA   MOROCCO (SPANISH ZONE). Servicio de Estadística.
    Estadística del comercio exterior en la Zona del Protectorado Español.

    1949-55   382.642
                         M867e

SPANISH WEST AFRICA

SA   SPANISH WEST AFRICA. Secretaría General.
    Sahara Español; anuario estadístico.

    1950   316.48
                        S735

SUDAN

SA   SUDAN. Ministry of Finance and Economics.
    Economic survey. 1956-

    1962-64+   330.9626
                         S943
    HL 1961   HA2275
                         S8A3

SA   SUDAN. Governor-General.
    Report on the administration, finances and conditions of the Sudan.
    Issued in Gt. Brit. Parliament. Papers by command; formerly published in Reports on the finances ... of Egypt

(continued)

and the Soudan, issued by Gt. Brit. High Commissioner for Egypt and the Sudan.

1921-49
Reports for 1939-41 and Brit. Docs.
   1942-44 issued in 1950.

HL   1939-47   HC537
                              S8A4

SB   SUDAN. Dept. of Statistics.
      Internal statistics. 1960/61-

SB   SUDAN. Wizārat al-Māliyah wal-Iqtiṣād.
      Sudan economic and financial review.
Jan.-June 1961-

HL   1961   HA2275
                              S8A4

CA   SUDAN. Dept. of Statistics.
      Foreign trade report, with some internal statistics. Annual.
      Issued 19 -50 by Dept. of Economics and Trade.
      Most tables include comparative statistics for the preceding five years.

1948-63+   382.9626
                            S943

CA   SUDAN. Dept. of Economics and Trade.
      Report.

no. 35, 1945   HF264
                            S7A27

CA **SUDAN. Customs Dept.
      Annual statistical returns of the Sudan customs. 1904-
      Through 1916 as Annual statement of foreign trade with foreign countries and Egypt.

CB   SUDAN. Dept. of Statistics.
      Foreign trade and internal statistics. Monthly.
      Supersedes its Monthly report and statistical return.
      Issued   -Nov. 1952 by the Dept. of Economics and Trade.

1957-Aug. 1965+   382.9626
                            S943f

SB **SUDAN. Customs Dept.
      Monthly statistical return of the Sudan customs. 1917.
      Continues Quarterly statement ....

CB **SUDAN. Customs Dept.
      Quarterly statement of trade with foreign countries and Egypt 1916//
      Continued as Monthly statistical return ....

AR   SUDAN. Ministry of Agriculture.
      Annual report.

FRI   1949-54

SUDAN, FRENCH

see also FRENCH WEST AFRICA
         MALI, 1960-

SB   SUDAN. Service Fédéral de la Statistique et de la Mécanographie.
      Bulletin statistique et économique mensuel.

SB   SUDAN. Service Statistique.
      Bulletin statistique mensuel.

CB   SUDAN. Service Fédéral de la Statistique et de la Mécanographie.
      Echanges commerciaux de l'ensemble Sénégal, Soudan, Mauritanie; commerce spécial.

SURINAM

SA   SURINAM. Algemeen Bureau voor de Statistiek.
      Jaarcijfers voor Suriname. Statistical yearbook of Surinam.
      Subseries of its Suriname in cijfers (Surinam in figures).

1956-60 (ser. no. 26),   HA1037
   1958-62 (ser. no. 34)+ S9A3

SA   NETHERLANDS (KINGDOM, 1815-   ).
       Departement van Overzeesche
       Gebiedsdeelen.
       Surinaamsch verslag. 1932-
       In two parts each year, 1932-   :
     I. Tekst van het verslag van bestuur en
     staat van Suriname over het jaar 1931- ;
     II. Statistisch jaaroverzicht van Suri-
     name over het jaar 1931-

     1937-38, 1944: pt. 1,   325.3492
       1945: pt. 2, 1946:   N469s
       pt. 1-2, 1947: pt. 1-2,
       1948: pt. 1-2, 1949:
       pt. 1-2, 1952: pt. 2

SA **SURINAM.
       Surinaamsche almanak.... 1829-
       Title varies slightly.

SB   SURINAM. Algemeen Bureau voor de
       Statistiek.
       Statistische berichten. 1956-

     Sept. 1959-May 1966+   330.98422
                                   D975

CA   SURINAM. Algemeen Bureau voor de
       Statistiek.
       In- en uitvoer. Imports and exports.
       Issued in its series Suriname in
     cijfers (Surinam in figures).
       Continues Handelsstatistiek.

     1953-55, 1957, 1960-64+   382.8422
                                   S961

CB   SURINAM. Algemeen Bureau voor de
CA   Statistiek.
       Maandstatistiek van de in- en uitvoer
     per goederensoort en per land.
       Dec. issue gives annual cumulation.

AR   SURINAM. Departement van Landbouw,
AB   Veeteelt en Visserij.
       Surinaamse landbouw. 1, Jan./Feb.
     1953-
       Dutch with English summaries.
       Yearly survey given in a special issue.
     Annual.

       FRI 1963-64+
            Annual: 1964-65+

AR   SURINAM. Departement van Landbouw,
       Veeteelt en Visserij.
       Verslag.
       Reports for   -1907 issued by Inspec-
     tie van den Landbouw in West Indië;
     1908-19 by Departement van den Land-
     bouw in Suriname.
       Similar material for 1955(?)-63 to be
     found under: Paramaribo, Surinam.
     Landbouwproefstation. Verslag (S201
     A52).
       Annual reports published as special
     issues under the series De Surinaamse
     landbouw, from 1964.

     1909-14, 1916-20,   S201
       1950-54   A5

SWAZILAND

SA   GREAT BRITAIN. Office of Common-
       wealth Relations.
       Report on Swaziland. 1946-
       Previously issued by Colonial Office in
     its numbered series, Colonial reports--
     annual (325.342 G787) no. 1102: 1920-
     no. 1921: 1938. Suspended 1940-46.

     1946-63+   325.342
                                   G7871Sw

SA **SWAZILAND.
       Blue book. 1913/14-

CA   SWAZILAND.
       Swaziland commerce report.

     1960/61

AR   SWAZILAND. Dept. of Agriculture.
       Annual report.

       FRI 1962-64+

AR   SWAZILAND. Dept. of Land Utilization.
       Report.

       FRI 1960-62

SWEDEN

SA SWEDEN. Statistiska Centralbyran.
    Statistisk arsbok för Sverige. 1,
    1914-
    Continues Sveriges officiella statistik
    i sammandrag.

    1915, 1918-33, 1935-39,   HA1523
    1941-52, 1954, 1956-65+   A46

SA SWEDEN. Statistiska Centralbyran.
    Historisk statistik för Sverige. His-
    torical statistics of Sweden.
    Began publication in 1955.

    no. 1-3+   HA1525
                                   A52

SA **SWEDEN. Statistiska Centralbryan.
    Sveriges officiella statistik i sam-
    mandrag. 1-44, 1870-1913//
    Vol. 1 published separately; 2-44,
    1871-1913, as no. 1 of its Statistisk
    tidskrift. Continued as Statistisk arsbok
    för Sverige.

SA SWEDEN. Statistiska Centralbyran.
    Sveriges officiella statistik.... 1911-
    Continues Bidrag till Sveriges offi-
    ciella statistik.
    Consists of subseries, or groups,
    catalogued separately.

SB SWEDEN. Statistiska Centralbyran.
    Allmän manads statistik. Monthly
    digest of Swedish statistics. 1, Jan.
    1963-

    Jan. 1963-Apr. 1966+   314.85
                                   A97a

SB SWEDEN. Statistiska Centralbyran.
    Statistisk tidskrift. 1, 1952-

    1952-66, no. 1+   314.85
                                   S797

SB **SWEDEN. Statistiska Centralbyran.
    Statistisk tidskrift. 1-154, 1860-
    1919//
    A statistical abstract. Sveriges offi-
    ciella statistik i sammandrag is included
    annually. 1895-1903 include also statis-
    tical tables of foreign countries, Statis-
    tiska öfversigtstabeller för olika länder.
    Through 1892 an account of the banks of
    Sweden is included, and is continued in
    Bidrag till Sveriges officiella statistik,
    letter Y.

SB SWEDEN. Statistiska Centralbyran.
SA   Statistiska meddelanden. Statistical
    reports.
    Specialized statistics issued in various
    subseries.

CA SWEDEN. Statistiska Centralbyran.
    Utrikeshandel.
    Began publication with the volume for
    1911.
    1911-59, del 1, issued by Kommer-
    skollegium; 1959, del 2- by Statistiska
    Centralbyran.
    Title varies: 1911-61, Handel; berät-
    telse.

    1914-15, 1918-64: 1-2+   382.485
                                   S974

CB SWEDEN. Statistiska Centralbyran.
    Manadsstatistik över utrikeshandeln.
    Began publication in 1913.
    1913-61 issued by Kommerskollegium
    (1952-61 as Bilaga to its Kommersiella
    meddelanden).
    Issued in series Sveriges officiella
    statistik: handel.

    1950-Feb. 1966+   382.485
                                   S974m

CB SWEDEN. Statistiska Centralbyran.
    Kvartalsstatistik över utrikeshandeln.
    1961-
    Issued in series Sveriges officiella
    statistik: handel.
    In two parts every third month: pt. 1,
    Införsel; pt. 2, Utförsel.

    Jan. 1965-Dec. 1965+   HF217
                                   A4

AA   SWEDEN. Statistiska Centralbyran.
      Jordbruksstatistisk arsbok. Statistical yearbook of agriculture. 1965-

      FRI   1965+

AA   SWEDEN. Statistiska Centralbyran.
      Jordbruk och boskapsskötsel.

      1909-63+   630.6485
                                        S974a

      HL   1914-19

AA   SWEDEN. Statistiska Centralbyran.
      Arsväxten ... preliminär redogörelse.
      1911-

      1914-22, 1924-36, 1944-62+   630.6485
      wanting: 1933   S974ar

SWITZERLAND

SA   SWITZERLAND. Statistisches Amt.
      Statistisches Jahrbuch der Schweiz
      .... Annuaire statistique de la Suisse
      1891-
        1891-1919 is Schweizerische Statistik .... In place of the yearbook for 1897 there was issued the Graphisch-statistischer Atlas der Schweiz ... 1897. Atlas graphique et statistique [etc.], and that publication is included in the index. Index: 1-10, 1891-1901.

      1910-20, 1924, 1926-63+   HA1593
                                        A4

      HL   1921-25, 1939-44,
            1950

SA   SWITZERLAND.
      Staats-Kalender der Schweizerischen Eidgenossenschaft .... Annuaire de la Confédération suisse .... Annuario della Confederazione svizzera. 1849-
        1849-1867/68 in both French and German editions. Since 1918 the Italian section has been added.

SA **SWITZERLAND. Statistisches Bureau.
      Schweizerische Statistik 1-217, 1860-1919//
        In German, French, German and French, or French and Italian.
        Certain numbers of this set form the Statisches Jahrbuch der Schweiz ... Annuaire statistique de la Suisse, 1891- , which is entered separately.
        Superseded by Schweizerische statistische Mitteilungen.

SB **SWITZERLAND. Statistisches Bureau.
      Schweizerische statistische Mitteilungen .... Bulletin de statistique suisse .... Bollettino di statistica svizzera.
      1, 1919-
        Scattered numbers in both French and German.
        Supersedes its Schweizerische Statistik.

CA   SWITZERLAND. Oberzolldirektion.
      Jahresstatistik des Aussenhandels der Schweiz. Statistique annuelle du commerce extérieur de la Suisse.
        Issued in 3 vols. each year.

      1952-65, v. 1+   382.494
                                        S979j

CA   SWITZERLAND. Oberzolldirektion.
      Statistik du commerce suisse. Rapport annuel 1885-
        Title page and tables in German and French, 1885-86.
        Title varies: 1885-86 as Schweizerische Handelsstatistik. Tabelle der Einheitswerthe pro 1885-86 nebst Angabe der Ausfuhr-mittlewerthe von Deutschland, Frankreich und Italien pro 1884-85 ...; 1889-93, Statistique du commerce de la Suisse. Rapport annuel et tableau des valeurs moyennes ... (subtitle varies); 1897, Schweizerische Handelsstatistik. Jahresbericht.

      1910, 1912, 1914-23,   382.494
      1925-26, 1928-39: pt. 2,   S979a
      1952-64+

CA   SWITZERLAND. Finanz- und Zollde-
       partement.
     Statistik des Warenverkehrs der
Schweiz. Statistique du commerce de la
Suisse avec l'étranger. 1885-

     1910-27, pt. 1   382.494
                              S979

CA **SWITZERLAND. Oberzolldirektion.
     Uebersichts-tabelle der Ein-, Aus-
und Durchfuhr .... Tableau général de
l'importation, de l'exportation, et du
transit .... Prospetto generale delle
importazioni, exportazioni e del transito.
1839-84//
     1839-50 as Général tableau ....

CB   SWITZERLAND. Oberzolldirektion.
     Monatsstatistik des Aussenhandels der
Schweiz .... Statistique mensuelle du
commerce extérieur de la Suisse.
     Continues Statistik des auswärtigen
Handels.

     1925-26, Dec. 1948-   382.494
       Dec. 1965+   S979m
     wanting: May, Aug., Oct.,
       Dec. 1926
     1948- June and Dec.
       issues only

     FRI Current two years only.

CB   SWITZERLAND. Oberzolldirektion.
     Statistique semestrielle du commerce
extérieur de la Suisse avec les divers
pays. (?) -1964//

CB **SWITZERLAND. Oberzolldirektion.
     Statistik des auswärtigen Handels der
Schweiz .... Statistique du commerce
extérieur de la Suisse .... Quarterly.
1885-1927//
     Title varies: 1885-92 as Warenver-
kehr der Schweiz mit dem Auslande.
Uebersicht der Ein- und Ausfuhr der
wichtigsten Warenartikel ...; 1893-
1922, Schweizerische Handelsstatistik.
Ein- und Ausfuhr der wichtigsten Waren
... (subtitle varies).
     Continued as Monatsstatistik des aus-
wärtigen Handels der Schweiz.

AA   SWITZERLAND. Volkswirtschaftsde-
       partement.
     Annuaire agricole de la Suisse. 1,
1900-
     16, 1915- as "Annexe au bulletin de
l'office vétérinaire et de la division de
l'agriculture du département fédéral de
l'économie publique."

     FRI 1910-19, 1922,
            1924-25, 1932,
            1944-46

AA **SWITZERLAND. Volkwirtschaftsde-
       partement.
     Landwirtschaftliches Jahrbuch ....
1, 1887-
     1887-1914 issued by Handels-,
Industrie- und Landwirtschaftsdeparte-
ment.
     30- issued as supplement to
Veterinäramt ... Mitteilungen.

SYRIA

see also UNITED ARAB REPUBLIC

SA   SYRIA. Ministry of Planning. Directo-
       rate of Statistics.
     Statistical abstract.
     Began publication with volume for
1948.
     Issued 1948-54 by Syria. Ministry of
National Economy. Directorate of Statis-
tics; 1957-60 by United Arab Republic
(Syrian Region). Ministry of National
Planning. Directorate of Statistics.
     Issues for 1948-49 in Arabic only.

     1949-64+   315.69
                              S995

SA   GREAT BRITAIN. Dept. of Overseas
CA   Trade.
     ... Economic conditions in Syria ....
[1920-38.]
     Title varies: 1936/38- Report on
economic and commercial conditions in
Syria and the Lebanon ....

     1920, 1922, 1923, 1925,   380.005
       1928, 1930, 1932,   G786
       1934, 1936-38

SA   Conseil Supérieur des Intérêts Communs
         (Syria and Lebanon). Service des
         Etudes Economiques et Statistiques.
         Recueil de statistiques de la Syrie et
du Liban. 1942-

     HL   1942/43-1944,   HA1941
             1945-47, 1948-49   A3

SB   SYRIA. Ministry of Planning. Directo-
         rate of Statistics.
         General bulletin of current statistics.
1, Jan. 1951-

     v. 1, no. 1, Jan. 1951-   HA1941
     v. 10, 4th Q. 1960;   A33
     v. 12, no. 3, Sept.
     1962-Dec. 1964+

CA   SYRIA. Direction Générale des Douanes.
         Statistiques du commerce extérieur.

     1950-53, 1960-63+   382.569
                                     S995

     HL   1950

CA   Conseil Supérieur des Intérêts Communs
         (Syria and Lebanon). Administration
         des Douanes.
         Statistiques du commerce extérieur.

     1939/43: v. 1-2, 1944-47   HF259
                                           S95A3

CB   SYRIA. Ministry of Planning. Directo-
         rate of Statistics.
         Summary of foreign trade. 1951-
         Issue for the 4th quarter contains
figures for the whole year.

     Dec. 1951/52, June, Sept.,   382.569
     Dec. 1953, June, Sept.,   S996
     Dec. 1954, Dec. 1955-
     Dec. 1961, Jan./Sept.
     1962-Oct./Nov. 1965+

CB **SYRIA. Office Commercial pour la Syrie
         et le Liban.
         Bulletin économique. 1, 1921-

T

TAIWAN

see FORMOSA

TANGANYIKA

see also EAST AFRICA
            TANZANIA

SA   TANGANYIKA. Treasury. Economics
         and Statistics Division.
         Statistical abstract. 1961-
         Supersedes the Statistical abstract
issued by the Tanganyika Unit of the East
African Statistical Dept. of the East
Africa High Commission.

     1961-63+   HA2005
                                            T3A4

     FRI   1962-63+

SA   East Africa High Commission. East
         African Statistical Dept.
         Statistical abstract. Tanganyika.
1938/51-1960//
         Superseded by Tanganyika. Treasury.
Economics and Statistics Division. Statistical abstract, 1961-

     1938/52, 1955-58, 1960   316.78
                                           E13

     FRI   1938/52

SA   GREAT BRITAIN. Colonial Office.
         Report by His Majesty's government
in the United Kingdom of Great Britain
and Northern Ireland to the General
Assembly of the United Nations on the
administration of Tanganyika under
United Kingdom trusteeship. 1947-

     HL   1948-60   DT445
                                           A5G77

SA   GREAT BRITAIN. Colonial Office.
     ... Report by His Britannic Majesty's government to the Council of the League of Nations on the administration of Tanganyika Territory ....
     1920-21 issued as Papers by command; 1923 issued as Colonial Office. Colonial no. -

1921-22	Brit. Docs.
1923-60	325.342 G784
HL 1920-38, 1948-60 (in 2 pts.)	DT445 A5G77

SA   TANGANYIKA TERRITORY.
     Blue book. 1921-48//

1945-48	354.678 T164

CA   TANGANYIKA. Ministry of Commerce and Industry.
     Commerce and industry in Tanganyika.

1957, 1960, 1961	330.9678 T165

CA   TANGANYIKA. Customs Dept.
     Trade report. Annual. 1919/20-1948//
     For later see EAST AFRICA.

1946-48	382.678 T164

CB   TANGANYIKA. Ministry of Commerce and Industry.
     Tanganyika trade bulletin. no. 1-13, Apr./May 1955-May 1961.
     Superseded by Tanganyika trade journal (later called Tanzania trade journal).

no. 2-13, 1955-61	330.9678 T168

CB   TANGANYIKA. Customs Dept.
     Trade and information report.
  Monthly.
     For later see EAST AFRICA.

July 1933-Oct. 1937

AR   TANGANYIKA. Dept. of Agriculture and Cooperative Development.
     Annual report. 1922-
  1951- issued in two parts.

FRI   1945-62+

TANZANIA

see also   TANGANYIKA
             ZANZIBAR

SB   TANZANIA. Central Statistical Bureau.
     Monthly statistical bulletin. v. 14, 1964-
     Supersedes the Statistical bulletin published by the East African Statistical Dept. of Tanganyika and continues its volume numbering.

Jan. 1964-Feb. 1966+	HA1977 T4A3

CA   TANZANIA. Custom House.
     Trade report. 1962-

CB   TANZANIA. Ministry of Industries, Mineral Resources and Power.
     Tanzania trade journal. no. 1, July/Sept. 1962-
     Title varies: July/Sept. 1962-Jan./Mar. 1964, Tanganyika and Zanzibar trade journal.
     Supersedes Tanganyika trade bulletin. Includes statistical section.

no. 1-14, July/Sept. 1962- Oct./Dec. 1965+	HF3899 T35T3
wanting: no. 4	

CB   TANZANIA. Custom House.
     Monthly trade information.

THAILAND

SA   THAILAND. Central Statistical Office.
       Statistical year book, Thailand.
       Began publication with issue for 1916; suspended between 1944 and 1952.
       Agency and title vary.

       1916, 1918-19, 1928/29-   HA1781
       1945/55, 1952, 1956/58,   A3
       1963-64+

SA   THAILAND.
       Thailand official yearbook. 1964-

       1964+   DS586
                                           T5

SA   GREAT BRITAIN. Dept. of Overseas
CA   Trade.
       ... Report on economic and commercial conditions in Siam .... [1919-39.]
       Report for 1919 issued in the series of Parliamentary Papers as Papers by command. Cmd. 795.
       Title varies.

       1919-24, 1926, 1928, 1930,   380.005
       1932, 1934, 1937, 1939   G786

SB   THAILAND. Central Statistical Office.
       Bulletin of statistics. 1, June 1952-

       June 1952-Dec. 1965+   315.93
                                           T364

    FRI   Current two years only.

CA   THAILAND. Customs Dept.
       Statement of the foreign trade and navigation 1907/08-
       Supersedes its Statistics of the import and export trade of Siam. Port of Bangkok.
       Title and agency name vary.

       1913/14-1918/19, 1930/31,   382.593
       1941, 1947-64: v. 1-2+   T364

CA **SIAM. Dept. of Customs and Excise.
       Port of Bangkok ... Abstract of foreign trade.

CB   THAILAND. Customs Dept.
       Reports of the imports and exports of Thailand. Monthly.
       Began publication in 1954.

       Jan. 1954-June 1965+   382.593
                                           T364q

AA   THAILAND. Ministry of Agriculture.
       Agricultural statistics of Thailand.
       Title varies: 1954, Statistical review of Thai agriculture.

    FRI   1959-63+

TIMOR

SB   TIMOR, PORTUGUESE. Serviços de Administração Civil. Secção de Estatística e Informações.
       Boletim trimestral de estatística. Bulletin trimestriel de statistique. 1, 1956-

SB **TIMOR.
       Boletim oficial do distrito aurome de Timor. 1, Jan. 6, 1900-
       Continues title under Macau.

CA **TIMOR.
       Resumo estatístico do movimento comercial e aduaneiro. 1910.

CB **TIMOR.
AB   Boletim do comercio, agricultura e fomento 1-9, 1912/20//

TOBAGO

see also TRINIDAD AND TOBAGO

SA **TOBAGO.
       Blue book. 1809-88//

AA **TOBAGO. Botanic Station.
Annual reports.
1906/07-1907/08 in Trinidad. Botanical Dept. Annual report; 1908/09- in Dept. of Agriculture. Administrative report.

TOGO

SA TOGO. Service de la Statistique Générale du Togo.
Inventaire économique du Togo.

1958, 1959/61   330.96681
                            T645

SB TOGO. Service de la Statistique Générale.
Bulletin de statistique. 1, 1952-

1957-Jan. 1966+   316.681
wanting: Jan.-Aug. 1957,   T645
Oct.-Dec. 1960,
Sept.-Dec. 1961

TOGOLAND
(British Administration)

SA GREAT BRITAIN. Colonial Office.
Report by His Majesty's government in the United Kingdom of Great Britain and Northern Ireland to the General Assembly of the United Nations on the administration of Togoland under United Kingdom trusteeship. 1947-
Issued in its series Colonial no. -

1947-55   325.342
                            G784

SA GREAT BRITAIN. Colonial Office.
Report by His Majesty's government in the United Kingdom of Great Britain and Northern Ireland to the Council of the League of Nations on the administration of Togoland under British mandate. 1920/21-1937.
1922 classed separately (916.681 G786).
Report for 1920/21 issued in the parliamentary series as Papers by command. Cmd. 1698 (Brit. Docs. C1922, v. 16).

(continued)

Reports for 1923- issued in the Office's series of papers called Colonial no. -

1920-37   325.342
                            G784

SA GREAT BRITAIN. Colonial Office.
West Africa. Report on the British mandated sphere of Togoland for ....
1920-55.

1920/21 (Cmd. 1698)   Brit. Docs.
                            C1922
                            v. 16

1922; 1923-55   916.681
                            G786

HL 1924   DT582
                            A3a
                            1924

TOGOLAND
(French Administration)

SA FRANCE.
Annuaire statistique de l'Afrique occidentale française et du territoire du Togo sous mandat de la France.

SA FRANCE.
Rapport annuel du gouvernement français a l'Assemblée Générale des Nations Unies sur l'administration du Togo placé sous la tutelle de la France. 1947-

1947   325.344
                            F811t
                            f

HL 1947, 1951-53   DT582
    (microfilm)   A5A65

SA TOGOLAND.
Rapport annuel du gouvernement français sur l'administration sous mandat des territoires du Togo.

HL 1921-38   DT582
                            A5

SB FRANCE. Service de la Statistique Générale.
Bulletin statistique mensuel du Togo.

TONGA

SA GREAT BRITAIN. Colonial Office.
    Annual report on Tonga. 1946-
    Previously issued in its numbered series, Colonial reports--annual (325.342 G787) no. 1291: 1924-25-- no. 1918: 1938. Suspended 1940-46.
    Earlier reports issued by the Foreign Office, 1909-1914/15, 1923/24. None issued 1915/16-1922/23//(?)

    1946-61+   325.342
                                G7871To

CA TONGA. Dept. of Customs.
    Statement of trade and navigation.
    Publisher states this is not available for distribution.

CA **TONGA.
    Trade reports. 1923.

AR TONGA ISLAND. Dept. of Agriculture.
    Report.

    1951-63+   630.69612
    wanting: 1961   T665

TRANSJORDAN

see JORDAN
    PALESTINE

TRIESTE
(British-U.S. Zone)

SA TRIESTE (Free Territory, British-U.S. Zone). Allied Military Government.
    Report of the administration of the British-U.S. Zone of the Free Territory of Trieste.

    no. 1-11, 1947-51   943.6
                                T824

SB TRIESTE (Free Territory, British-U.S. Zone). Allied Military Government.
    Bollettino di statistica della Zona anglo-americana del Territorio Libero di Trieste. 1946-
    Bimonthly. No longer published.

TRINIDAD AND TOBAGO

see also TOBAGO

SA TRINIDAD. Central Statistical Office.
    Statistical digest.
    Began publication with the issue for 1935/51.

    1935/51-1963+   HA867
                                A35

SA GREAT BRITAIN. Colonial Office.
    Annual report on Trinidad and Tobago.
    Previously issued in its numbered series, Colonial reports--annual (325.342 G787) no. 1103: 1920- no. 1915: 1938. Suspended 1940-46.

    1946-57+   325.342
                                G7871Tr

SA TRINIDAD.
    Blue book.
    Began publication in 1821. Volumes through 1887 include only Trinidad.

    1913/14-1922   354.7298
                                T833b

SA **TRINIDAD.
    Year book. 1, 1866-

SB TRINIDAD. Central Statistical Office.
    Economic report. Quarterly.
    June/Sept. 1950-

    Oct./Dec. 1950-   330.97298
      Oct.-Dec. 1965+   T832
    wanting: Apr./Sept. 1951

CA TRINIDAD. Central Statistical Office.
    Overseas trade report. 1, 1951-
    Earlier reports were compiled by the Cutoms Dept. and published as Council papers in Trinidad and Tobago. Legislative Council. Minutes of proceedings and Council papers.

    1952-63+   382.7298
                                       T832

CA TRINIDAD. Customs and Excise Dept.
    Administration report of the comptroller.

    1931, 1933-36, 1949-50,   382.7298
        1952-54, 1956-57, 1963+   T833

CB TRINIDAD. Central Statistical Office.
    Overseas trade; monthly report.
    v. [1], 1951-

    Sept. 1964-Dec. 1965+   HF157
    wanting: Dec. 1964   T8A46

AA **TRINIDAD AND TOBAGO. Botanical Dept.
    Annual report. 1879-1907/08//
    Continued in Dept. of Agriculture. Annual report.

AA **TRINIDAD. Government Statistician.
CA   Trade and crop returns.

AB **TRINIDAD AND TOBAGO. Dept. of Agriculture.
    Agricultural notes. Monthly. 1, 1930-

AB **TRINIDAD AND TOBAGO. Dept. of Agriculture.
    Bulletin. no. 1-21, pt. 2, 1888-1927//
    1-8, 1888-1908 as its Botanical Dept. Bulletin of miscellaneous information (title varies slightly).
    Suspended 1923-24.

AR TRINIDAD. Dept. of Agriculture.
    Administration report of the director of agriculture.
    Issued as Trinidad. Legislative Council. Council paper, no. 155-
    Title varies.

    1909/10, 1911, 1913/14,   630.67298
        1916-18, 1922-25   T833

TUNISIA

SA TUNISIA. Secretariat of State for Plan and Finance.
    Economic yearbook of Tunisia (Union Tunisienne du Commerce et de l'Industrie). 1964-

    HL v. 1, 1964+   HC547
                                       T9I91

SA TUNIS. Service des Statistiques.
    Annuaire statistique de la Tunisie. 1940-
    Supersedes Tunis. Direction des Affaires Economiques. Statistique générale de la Tunisie.

    1940-1961/62+   HA2071
                                       T52

SA GREAT BRITAIN. Dept. of Overseas
CA   Trade.
    ... Report on economic conditions in Algeria, Tunisia and Tripolitania ....
    [1921-36.]
    Title varies.

    1921-23, 1929-30, 1932,   380.005
        1935, 1936, 1937   G786

SA TUNIS. Direction des Affaires Economiques.
    Statistique générale de la Tunisie. 1913-39.
    Continues Rapport au président de la République sur la situation de la Tunisie. Statistique générale de la Tunisie, 1881/92-1912.
    Superseded by Tunis. Service des Statistiques. Annuaire statistique de la Tunisie.

    1914-22   316.11
                                       T926

SB   TUNISIA. Service Tunisien des Statistiques.
       Bulletin mensuel de statistique. June 1954-

SB   TUNISIA. Service Tunisien des Statistiques.
       Bulletin du Service Tunisien des Statistiques. 1947-
       Quarterly; irregular.

CA   TUNISIA. Direction des Douanes.
       Documents statistiques sur le commerce de la Tunisie. 1885-

| 1906, 1909-12, 1914-22 | 380.09611 T926 |

CB   **TUNIS. Direction des Douanes.
       Bulletin comparatif trimestriel du mouvement commercial de la Tunisie. 1928(?)-1940//
       Each issue cumulates from Jan. 1 and gives data for the corresponding period of the preceding two years.

TURKEY

SA   TURKEY. Istatistik Umum Müdürlügü.
       Annuaire statistique. 1928-

| 1930, 1932/33-1936/37, 1942/43, 1942/45, 1948-53, 1957-63+ | HA1911 A3 |

SA   TURKEY. Istatistik Umum Müdürlügü.
       Küçük istatistik yilligi. Statistical abstract. 1937/38(?)-

| 1937/38-1946, 1948, 1947-50, 1951, 1954 (no more published) | 314.96 T939s |

SA   GREAT BRITAIN. Dept. of Overseas
CA   Trade.
       ... Economic conditions in Turkey ... Report .... [1919-39.]
       1919 issued in the series of Parliamentary Papers as Parliament. Papers by command. Cmd. 942.
       Title varies.

| 1919, 1921, 1922, 1924, 1925, 1927, 1928, 1930, 1932, 1934, 1936, 1939 | 380.005 G786 |

SA   **TURKEY. Istatistik Müdüriyeti Umumiyesi.
       Publications .... no. 1, 1927-
       No. 1-2 in Turkish only; no. 3- in two editions, Turkish and French.
       Contents: 1. Loi de statistique; 2. Resultats du recensement générale de la population ... 1927; 3. Population de la Turquie par villayets et cazas, par villes et villages ... 1927; 4. Annuaire statistique; 5. Rencensement générale de la population; etc.

SA   OTTOMAN EMPIRE. Ministère des Finances.
       Bulletin annuel de statistique. 1325:1 (1909:1)-

1326 (1910), 1327 (1911)

SA   [TURKEY. Ministry of Finance (?)]
       Recettes et dépenses générales, 1302-1326 [1885-1910] Ecole de finances; établissements financières: .... N.p. [1910?]
       290-325 p.
       Detached (?) from a statistical publication.

[1885-1910] bound with SA 1910, 1911

SB   TURKEY. Istatistik Umum Müdürlügü.
       Istatistik bülteni. Bulletin of statistics. 1, 1952-

| 1912-16, no. 23-94 (numbering stops), Dec. 1955-Dec. 1965+ | 315.6 T937i |

CA TURKEY. Istatistik Umum Müdürlügü.
CB   Dis ticaret aylik istatistik, özel
    ticaret. Statistique mensuelle du com-
    merce extérieur, commerce spécial.
        Each Dec. issue includes pt. 1 of
    annual statistics (Dis ticaret yillik
    istatistik) for the year; pt. 2-4 issued
    separately.
        1941-45 have title: Harici ticaret.

    1926-30, 1932, 1941-63:   382.496
        pt. 2+   T939
    (Monthly foreign statistics
        suspended 1961-63)

CA **TURKEY. Gümrükler Umum Müdürlügü.
        Harici ticaret istatistikleri ....
    Statistique du commerce extérieur,
    publié par la Direction générale des con-
    tributions directes. 1907/08-
        Title varies.

CB TURKEY. Devlet Istatistik Enstitusu.
        Aylik dis ticaret istatistikleri.
    Monthly foreign trade statistics.

    Apr./June 1964+

CB TURKEY. Devlet Istatistik Enstitusu.
        Dis ticaret ... Ozel ticaretmadde ve
    madde gruplari itibarile ithalat ve
    ihracat. Statistique du commerce exté-
    rieur ... Commerce spécial. Importa-
    tions et exportations par articles et par
    groupe d'articles. Jan./Mar. 1952-

    Included with its Dis ticaret
        aylik istatistik (382.496
        T939)

CB **TURKEY. Gümrükler Umum Müdürlügü.
        Bulletin de statistique du commerce
    extérieur de l'Empire Ottoman.

AA TURKEY. Devlet Istatistik Enstitusu.
        Zarai istatistik ozetleri. The sum-
    mary of agricultural statistics.

        FRI 1936-56, 1941-62,
             1942-64+

AR TURKEY. Istatistik Umum Müdürlügü.
        Zerai bünye ve istiksal. Agricultural
    structure and production.

    1934/50, 1946/53,   630.6496
        1946/54, 1954/58,   T939
        1958/60, 1959/61,
        1960/62+

TURKS AND CAICOS ISLANDS

SA GREAT BRITAIN. Colonial Office.
        Annual report on the Turks and Caicos
    Islands. 1946-
        Previously issued in its numbered
    series, Colonial reports--annual
    (325.342 G787) no. 1104: 1920-
    no. 1927: 1938. Suspended 1940-46.

    1946-62+   325.342
                                 G7871Tu

U

UBANGI-SHARI

see CENTRAL AFRICAN REPUBLIC

UGANDA

see also EAST AFRICA

SA UGANDA. Statistics Division.
        Statistical abstract. 1961-
        Supersedes the Statistical abstract
    issued by the Uganda Unit of the East
    African Statistical Dept. of the East
    Africa High Commission.

    1961-65+   HA2007
                                 U2A4

    HL 1962, 1963   HA2007
                                 U2A31

        FRI 1961-65+

SA   GREAT BRITAIN. Colonial Office.
         Report on Uganda. 1946-
         Previously issued in its numbered
      series, Colonial reports--annual
      (325.342 G787) no. 1112: 1920-
      no. 1903: 1938. Suspended 1940-46.

         1946--1962-63+   325.342
                                  G7871U

SA   East Africa High Commission. East
         African Statistical Dept. Uganda Unit.
         Statistical abstract [of Uganda].
      1957-60//
         Superseded by the Statistical abstract
      issued by the Statistics Division of the
      government of Uganda.

         1957-60   316.761
                                  E13

         FRI   1957-60

SA   UGANDA.
         Annual reports of the Kingdom of
      Buganda, Eastern Province, Western
      Province, Northern Province.

         HL   1939/46, 1948/60

SA   UGANDA (PROTECTORATE).
         Blue book.

         1914-28   354.676
                                  U26

SB   UGANDA. Ministry of Planning and
         Community Development.
         Economic and statistical bulletin.
      Monthly.
         Issued as a supplement to Uganda
      gazette.

         Oct. 1964-Apr. 1965+   HA2007
                                  U2A35

CA   UGANDA. Ministry of Commerce and
         Industry.
         Report. 1958/59-
         Continues its Dept. of Trade. Report.
         Title varies: 1958/59, A review of
      the activities.
         1961/62 not published.

         1958/59-1963/64+   382.676
                                  U28

CA   UGANDA. Ministry of Economic Affairs.
         The external trade of Uganda, 1950-
      1960.

         1950-60 (1 vol.)   382.678
                                  U29

         FRI   1950-60 (1 vol.)

CA   UGANDA. Dept. of Trade.
         Report.
         Began publication with Report for
      July/Dec. 1955; superseded by Ministry
      of Commerce and Industry. Report.

         1955-57   382.676
                                  U27

CA   UGANDA. Dept. of Commerce.
         Report. Annual.

         1951-54   382.676
                                  U26

AR   UGANDA. Dept. of Agriculture.
         Annual report. 1910/11-

         FRI   1914-23, 1949-63+

UNION OF SOUTH AFRICA

see SOUTH AFRICA

UNITED ARAB REPUBLIC

see also EGYPT

SA   UNITED ARAB REPUBLIC. Central
         Agency for Public Mobilisation and
         Statistics.
         Statistical handbook, United Arab
      Republic.

         1952-64 (1 vol.)   HA2042
                                  A35

SA UNITED ARAB REPUBLIC. Central
    Agency for Public Mobilisation and
    Statistics.
    <u>Statistical pocket-book.</u>
    Also published in Arabic and French.
    Issued 19   by the Ministry of Finance
and Economy, Statistical Dept., of Egypt;
19 -62 by Dept. of Statistics and Census
of the U.A.R. (called, 19 -59, Statistical Dept.).
    Title varies.

    1953-63+   HA2042
    wanting: 1955   A4

    HL   1959-62+   HA2042
                                     A4

SA UNITED ARAB REPUBLIC. Central
    Agency for Public Moblisation and
    Statistics.
    <u>Basic statistics.</u> 1962-
    Series in Arabic began Jan. 1960;
English version, June 1962.

    1963+   HA2042
                                     A36

SA UNITED ARAB REPUBLIC. Dept. of
    Statistics and Census.
    <u>Ten years of revolution; statistical
atlas.</u> 1952-62 (1 vol.).

    1952-62   HC535
                                     A5685

SA UNITED ARAB REPUBLIC. Central
    Agency for Public Mobilisation and
    Statistics.
    <u>Annuaire statistique.</u> 1909-
    Issuing agency varies. Earlier vol-
umes issued by Egypt. Ministry of
Finance and Economy.
    1962 in Arabic only.

    1910-14, 1916, 1919-   316.2
      1924/25, 1945/46-   E32
      1946/47, 1954/55-
      1955/56, 1960, 1961+

    HL   1962 (Arabic only)

CA UNITED ARAB REPUBLIC. Central
    Agency for Public Mobilisation and
    Statistics.
    <u>Annual bulletin of foreign trade.</u>
    1910-
    1910-17, French and Arabic; 1918-
English and Arabic.
    Title and issuing agency vary.

    1910-17, 1919-39,   382.62
      1944-49, 1959-61,   E32
      1964/65+

    HL   1962/63 (Arabic only)

CA UNITED ARAB REPUBLIC. Central
    Agency for Public Mobilisation and
    Statistics.
    <u>U.A.R. foreign trade, according to
standard international trade classifica-
tion (S.I.T.C.).</u> Revised.
    Title varies: 19 -52, <u>Egypt trade
statistics</u>; 1953- <u>Yearbook of international trade statistics</u>; 1960-61, <u>Statisti-
cal year book of Egypt's foreign trade.</u>

    1944-50, 1953-54,   HF263
      1960-61, 1963/64+   A46

CA UNITED ARAB REPUBLIC. Administra-
    tion des Douanes.
    <u>Rapport sur la commerce extérieur.</u>

    1959-62+   HF263
                                     A5

CB UNITED ARAB REPUBLIC. Central
    Agency for Public Mobilisation and
    Statistics.
    <u>Monthly bulletin of foreign trade.</u>
    English and Arabic.
    Title varies: 1918-63, <u>Monthly sum-
mary of foreign trade.</u>

    Jan. 1918-Jan. 1919,   382.62
      Jan. 1922-Dec. 1952,   E32m
      Jan. 1961-Dec. 1965+
    wanting: Apr., May 1918

AA UNITED ARAB REPUBLIC. Ministry of Agriculture.
    <u>UAR agriculture.</u>
    Title varies: <u>Agricultural statistics.</u>

    1965   S341
                                        A4

    HL 1950: v. 1-2 (Arabic only)

AB UNITED ARAB REPUBLIC. Central Agency for Public Mobilisation and Statistics.
    <u>Bulletin of agricultural and economic statistics.</u>
    Supersedes Egypt. Ministry of Finance. Statistical Dept. <u>... Monthly return ...</u>, 1912-19, and <u>... Monthly agricultural and economic statistics,</u> 1919-(?), various other agencies following.

    1912-51, Oct. 1958-   630.662
    Apr. 1963+   E32m
    wanting: several issues   1912-19 and
                                 1919-

UPPER VOLTA

SA UPPER VOLTA. Direction de la Statis-
SB   tique et des Etudes Economiques.
    <u>Bulletin de statistique.</u>
    Began publication in 1960. Issued by the agency under variant names: 1960, Bureau des Etudes Economiques et de la Statistique; 1961, Service de la Statistique.
    Its <u>Statistique annuelle</u> contained in two semiannual supplements, 1962-

    Sept. 1960-Dec. 1965+   315.625
                                 U68

CA IVORY COAST. Service de la Statistique Générale.
    <u>Statistiques du commerce extérieur de la Haute-Volta: importations, exportations, commerce spécial.</u>

    1955, 1957   382.6668
                                 I96eh

AR UPPER VOLTA. Direction des Services Agricoles.
    <u>Rapport annuel.</u> 1960-

URUGUAY

SA URUGUAY. Dirección General de Estadística.
    <u>Anuario estadístico.</u> 1884-
    Each year in two volumes.

    1894, 1902/03,   HA1071
    1907/08-1943   A3

SA URUGUAY. Dirección General de Estadística.
    <u>Síntesis estadística de la República del Uruguay.</u>

    1919-26   318.6
                                U82

SB URUGUAY. Dirección General de Estadística.
    <u>Boletín estadístico de la República Oriental del Uruguay.</u> Mar. 1903-May 1939.
    Semiannual, May and Nov.; no longer published.

SB URUGUAY. Dirección General de Estadística y Censos.
    <u>Boletín informativo.</u>

CA URUGUAY. Dirección General de Estadística.
    <u>Comercio exterior.</u>

    1915   382.86
                                U82

CA **URUGUAY. Dirección General de Estadística.
    <u>Comercio, navegación y hacienda pública. Estadística del comercio exterior ... Comercio interior. Existencia de ganados. Apuntes estadísticos.</u>
    1874/75//(?)

AA URUGUAY. Ministerio de Ganadería y Agricultura. Dirección de Agronomía. Sección Economía y Estadística Agraria.
    <u>Recopilación de la estadística agrícola del Uruguay.</u>

    FRI 1948, 1950

AA **URUGUAY. Dirección de Agronomía.
    Anuario de estadística agrícola.
    1892-
    Early years as Estadística agrícola.
    1892-94 issued by Ministerio de
    Fomento; 1895-1912 never published.
    1915- issued by various offices under
    Ministerio de Industrias.

AB URUGUAY. Departamento de Ganadería
    y Agricultura.
    Anales. 1-11, Jan. 1898-June 1908//
    Continues Comisión Central de Agri-
    cultura. Boletín oficial.

AB **URUGUAY. Comisión Central de Agri-
    cultura.
    Boletín oficial de la comisión.
    Continued as Departamento de Gana-
    dería y Agricultura. Anales.

V

VENEZUELA

SA VENEZUELA. Dirección General de
    Estadística.
    Anuario estadística de Venezuela.
    1877-1912, 1938-
    None published 1913-37.

    1908-12, 1944-52,   HA1091
    1954-1957/63   A4

    FRI 1945-51, 1954

SA GREAT BRITAIN. Dept. of Overseas
CA   Trade.
    ... Economic conditions in Venezuela
    ... Report .... [1921-35.]

    1921-23, 1925, 1927,   380.005
    1930, 1932, 1935   G786

SB VENEZUELA. Dirección General de
    Estadística.
    Boletín de estadística. Jan. 1941-

    Apr. 1942-Dec. 1944,   318.7
    Jan. 1946-June 1947,   V458b
    Jan. 1949-July 1965+
    wanting: several issues

    FRI 1949-65+

SB **VENEZUELA. Dirección General de
    Estadística.
    Boletín de estadística. 1-6 (no. 1-
    60), July 1904-June 1909//
    Continued by the Anuario estadístico
    de Venezuela and Boletín del Ministerio
    de Fomento.

CA VENEZUELA. Dirección General de
    Estadística.
    Estadística mercantil y marítima.
    Began publication with issue for
    1873/74; suspended 1876/77-1901-02.
    At head of title: 1873-76, Ministerio
    de Fomento; 1902/03, Ministerio de
    Hacienda.
    Agency and title vary.

    1919^2, 1923-24, 1926,   382.87
    1937, 1939-40, 1943-46,   V458
    1948-55 (no more pub-
    lished)
    wanting: 1954

    FRI 1955

CA **VENEZUELA. Dirección General de
    Estadística.
    Memoria 1874//
    Relates to statistics of commerce,
    navigation, and customs, 1870-73.

CB VENEZUELA. Dirección de Comercio
    Exterior y Consulados.
    Comercio exterior de Venezuela. 1,
    Jan. 1962-

    Dec. 1963-Jan. 1966+
    wanting: several issues

CB VENEZUELA. Dirección General de
    Estadística y Censos Nacionales.
    Boletín de comercio exterior. 1,
    Jan. 1959-

    Jan. 1959-Feb. 1965+   382.87
                                V4585

    FRI Current two years
         only.

AA VENEZUELA. Dirección de Economía y
    Estadística Agropecuario.
    Anuario estadístico agropecuario. 1,
    1961-

    FRI 1962, 1964+

AA VENEZUELA. Ministerio de Agricultura
    y Cria.
    Superficie, producción, rendimiento,
    importación y exportación de productos
    agropecuarios.

    FRI 1945-63 (1 vol.)

AA **VENEZUELA. Ministerio de Agricul-
CA   tura, Industria y Comercio.
      Memoria. 1, 1898//

AR VENEZUELA. Ministerio de Agricultura
    y Cria.
    Memoria y cuenta. 1936-
    In 1 vol. each year, 1936-1936/37;
    in 3 vols. each year 1937-(?)
    Earlier statistics are found in the
    Memoria del Ministerio de Salubridad y
    de Agricultura y Cria.

    1938-39: 1-2, 1960, 1962   630.987
                                V458

VIETNAM

see also INDOCHINA

SA VIETNAM. National Institute of Statis-
      tics.
    Statistical yearbook of Viet-Nam.
    1949/50-
    Earlier statistics for Vietnam were
    included in the Annuaire statistique
    issued by the Service de la Statistique
    Générale de l'Indochine.

    1949/50-1962+   HA173
                                A2

    FRI 1956

SB VIETNAM. National Institute of Statis-
      tics.
    Monthly statistical bulletin of Viet-
    Nam.
    Cover title in Annamese; running
    titles and text in Annamese, French, and
    English.

    Jan. 1957-Jan. 1966+   330.9596
                                V660

    FRI Current year only.

SB VIETNAM. Institut de la Statistique et
      des Etudes Economiques.
    Viet-Nam kink-te tap-san. Bulletin
    économique du Viet-Nam. Jan. 1950-
    Monthly.

CA VIETNAM. Direction Générale des
      Douanes.
    Statistiques du commerce extérieur du
    Viet-Nam. 1955-
    Vietnamese and French.
    Continues in part its Tableau général
    du commerce extérieur de l'union
    douanière des états du Cambodge, du
    Laos et Vietnam.

    1960-63+   382.596
                                V666

AA VIETNAM. Agricultural Economics and
      Statistics Service.
    Agricultural statistics yearbook.

    FRI 1963-64+

VIETNAM (Democratic Republic, 1946–)

SB VIETNAM (DEMOCRATIC REPUBLIC,
      1946-). Nha Kinh-te. Bac-viet
      Thong-ke Nguyet-san.
    Bulletin statistique mensuel du Nord-
    Vietnam. no. 1, 1954-

VIRGIN ISLANDS

SA GREAT BRITAIN. Colonial Office.
    British Virgin Islands; report.
    1955-56--
       Continues in part its Report on the Leeward Islands, issued in the series Colonial annual reports (325.342 G7871Le).

    1957--1961-62+   325.342
                                  G7871V

SA **VIRGIN ISLANDS.
    Blue book. 1821-96//(?)

AR VIRGIN ISLANDS (PRESIDENCY). Dept. of Agriculture.
    Report. 1906/07-

    1964+   S183
                                  V5A3

W

WEST INDIES

AA WEST INDIES. Federal Statistical Office.
    Agricultural statistics, series 1.
    (1956-58 survey)

    FRI 1956-58, no. 1-7

WESTERN SAMOA

SA NEW ZEALAND. Dept. of Island Territories.
    Western Samoa administered under trusteeship agreement dated 13th December 1946; report.

    HL   1949-60   DU819
         wanting: 1954, 1955   A2A20

CA WESTERN SAMOA.
    Trade, commerce and shipping for the territory of Western Samoa for the year.

    1925, 1928, 1941, 1944-51,   382.961
    1953-60, 1964+   S191

Y

YUGOSLAVIA

In addition to Serbia and Montenegro, Yugoslavia includes Dalmatia, Slavonia, Croatia, Bosnia and Herzegovina. Publications of these areas do not appear in this list separately.

SA YUGOSLAVIA. Savezni Zavod za Statistiku i Evidenciju.
    Statisticki godisnjak. 1929-
       Title and agency vary.
       Statistical yearbook, English text, issued as supplement, 1954-

    1936, 1938/39, 1940,   HA1631
       1954-64+   A336
    wanting: 1957

    HL   1955-64+   HA1631
                                  A34

    FRI 1940, 1954, 1956,
         1959

SA YUGOSLAVIA. Federal Institute for Statistics.
    Statistical pocketbook of Yugoslavia.
    1955-

    HL   1964   HA1631
                                  A352

    FRI 1958

SA YUGOSLAVIA. Savezni Zavod za Statistiku.
    Jugoslavija 1945-1964; statisticki pregled.

    HL   1945-64 (1 vol.)   HA1631
                                  A338

SA   YUGOSLAVIA. Savezni Zavod za
         Statistiku.
         <u>Jugoslavija izmedu VII i VIII kongresa
SKJ, statisticki podaci 1958-1964</u>.

      HL   1958-64 (1 vol.)   HA1631
                                     A33

SA   GREAT BRITAIN. Dept. of Overseas
CA   Trade.
         ... <u>Report on the economic and
industrial conditions of the Serb-Croat-
Slovene kingdom</u> .... [1921-38.]
         Title varies: 1930- <u>Economic condi-
tions in Yugoslavia</u> ....

      1921-25, 1928, 1930,   380.005
         1932, 1934, 1936, 1938   G786

SB   YUGOSLAVIA. Savezni Zavod za
         Statistiku i Evidenciju.
         <u>Statisticka revija</u>. Statistical review.
      Quarterly. 1951-

      god 1, Mar. 1951-   HA37
         god 16, Mar. 1966+   Y83SA

CA   YUGOSLAVIA. Ministarstvo Spoljne
         Trgovine. Uprava za Statistiku i
         Evidenciju.
         <u>Statistika spoljne trgovine</u>. <u>Statistics
of foreign trade</u>. 1918/20-
         Tables in Serbian and English.

      1946-49, 1950-63+   382.497
                                     Y94

Z

ZAMBIA

see also RHODESIA, NORTHERN

SB   ZAMBIA. Central Statistical Office.
CB   <u>Monthly digest of statistics</u>. no. 1,
         Apr. 1964-
            Issued through Sept. 1964 by Northern
         Rhodesia. Central Statistical Office.
            Supersedes in part the <u>Monthly digest
of statistics of the Federation of Rhodesia
and Nyasaland</u>, published by the Central
African Statistical Office.

      May 1964-Mar. 1966+   HA1977
                                     Z3A4

CA   ZAMBIA. Central Statistical Office.
         <u>Annual statement of external trade</u>.
      1964-

      1964+   HF265
                                     Z3A4

CB   ZAMBIA. Central Statistical Office.
         <u>External trade statistics: summary of
external trade</u>.

      Jan.-June 1965+   HF265
                                     Z3A38

CB   ZAMBIA. Central Statistical Office.
         <u>External trade statistics, quarterly</u>.
         Published for only a short period,
      1964(?).
         None available from publisher.

AB   ZAMBIA. Ministry of Agriculture. Eco-
         nomics and Marketing Division.
         <u>Monthly economic bulletin</u>.

      FRI   Nov. 1964, Sept. 1965

AR   ZAMBIA. Central Statistical Office.
         <u>Agriculture production in Zambia;
production on non-African farms plus
sales of African grown crops</u>. 1964-

      1964+   HD9017
                                     Z3A4

ZANZIBAR

see also TANZANIA

SA   ZANZIBAR.
         <u>Summary digest of useful statistics,
1961</u>.

      HL   1961   HA2275
                                     Z2A3

SA GREAT BRITAIN. Colonial Office.
    Annual report on Zanzibar. 1946-
    Previously issued in its numbered
    series, Colonial reports--annual
    (325.342 G787) no. 1091: 1920-
    no. 1892: 1938. Suspended 1940-46.

    1946-60+   325.342
                                G7871Z

SA ZANZIBAR.
    Blue book. 1913-

    1947   354.6781
                                Z33

SA ZANZIBAR. Chief Secretary's Office.
    Statistics of the Zanzibar Protectorate.
    Began publication with the volume for
    1893-1920.

    1895-1935   316.781
                                Z34

CA ZANZIBAR. Customs Dept.
    Annual trade report. 1920-

    1948: pt. 1-2, 1949-50,   382.6781
    1953: pt. 1-2, 1954-61+   Z34

CB ZANZIBAR. Customs Office.
    Monthly trade information. Jan.
    1946-
    First issue titled Trade and information report.
    No longer published.

AR ZANZIBAR. Dept. of Agriculture.
    Annual report. 1897-

    FRI 1957-60

HOOVER INSTITUTION BIBLIOGRAPHICAL SERIES

VIII. *The Chinese Communist Movement, 1921-1937*, by Chun-tu Hsueh. 1960. 131 p. $3.50.

X. *Guide to Russian Reference Books, Volume I: General Bibliographies and Reference Books*, by Karol Maichel. 1962. 92 p. Hard, $7.00; Paper, $5.00.

XI. *The Chinese Communist Movement, 1937-1949*, by Chun-tu Hsueh. 1962. 312 p. $5.00.

XIV. *United States and Canadian Publications on Africa in 1961*. Annual. 1963. 114 p. $3.00.

XV. *United States and Canadian Publications on Africa in 1962*. Annual. 1964. 104 p. $3.00.

XVI. *NSDAP Hauptarchiv: Guide to the Hoover Institution Microfilm Collection*, compiled by Grete Heinz and Agnes F. Peterson. 1964. 175 p. $4.50.

XVIII. *Guide to Russian Reference Books, Volume II: History, Auxiliary Historical Sciences, Ethnography, and Geography*, by Karol Maichel. 1964. 297 p. $16.00.

XIX. *German Africa*, by Jon Bridgman and David E. Clarke. 1965. 120 p. $8.00.

XX. *United States and Canadian Publications on Africa in 1963*. Annual. 1965. 136 p. $3.00.

XXI. *The Communist International and Its Front Organizations*, by Witold S. Sworakowski. 1965. 493 p. $15.00.

XXII. *Soviet Disarmament Policy, 1917-1963*, by Walter C. Clemens, Jr. 1965. 151 p. $4.00.

XXIII. *The Treason Trial in South Africa: A Guide to the Microfilm Record of the Trial*, by Thomas Karis. 1965. 124 p. $3.00.

XXIV. *Soviet and Russian Newspapers at the Hoover Institution: A Catalog*, compiled by Karol Maichel. 1966. 235 p. Hard, $6.50; Paper, $5.00.

XXV. *United States and Canadian Publications on Africa in 1964*. Annual. 1966. 180 p. $5.00.

XXVI. *Stalin's Works: An Annotated Bibliography*, compiled by Robert H. McNeal. 1967. 197 p. $5.00.

XXVII. *German Periodical Publications*, prepared by Gabor Erdelyi. 1967. 175 p. $5.00.

XXVIII. *Foreign Statistical Documents*, edited by Joyce Ball. 1967. 173 p. $5.00.

XXIX. *Handbook of American Resources for African Studies*, by Peter Duignan. 1967. 218 p. $6.00.

Unlisted volumes are out of print.

Orders should be sent to:
Publications Department
Hoover Institution
Stanford University
Stanford, California 94305